Survival

'Survival of the fittest is probably the best known reduction of Charles Darwin's thought, and this fascinating and accessible book examines the survival of the human race from a broad range of viewpoints. Through in-depth examinations of a number of very distinct aspects of human life, the book covers topics ranging from the preservation of Empires, to the challenges of maintaining cultural identity, the sufferings inflicted by famine, disease and natural disasters, the opportunities for increased longevity, and the threats presented by climate change. The chapters draw from the expertise of those in the arts and humanities, as well as the social, physical and biological sciences. Each chapter explores strategies which may be adopted to assist us in our individual struggle for existence and to preserve, and indeed improve, our collective lifestyles.

THE DARWIN COLLEGE LECTURES

Survival

THE SURVIVAL OF THE HUMAN RACE

Edited by *Emily Shuckburgh*

CAMBRIDGE
UNIVERSITY PRESS

CAMBRIDGE UNIVERSITY PRESS
Cambridge, New York, Melbourne, Madrid, Cape Town, Singapore, São Paulo

Cambridge University Press
The Edinburgh Building, Cambridge CB2 8RU, UK

Published in the United States of America by Cambridge University Press, New York

www.cambridge.org
Information on this title: www.cambridge.org/9780521710206

First published 2008

Printed in the United Kingdom at the University Press, Cambridge

A catalogue record for this publication is available from the British Library

ISBN 978–0–521–71020–6 paperback

Contents

Acknowledgements

A number of fellows and staff of Darwin College contributed to the lecture series and its publication. The editor would especially like to acknowledge the assistance of Joyce Graham at every stage of the process, and to convey warm thanks. The assistance of Grant Tapsell in the early stages of preparation of the lecture series, Rick Chen in obtaining copyright permissions for this book and Sarah King in providing transcripts of some of the lectures is also gratefully acknowledged.

1 Survival of the human race

EMILY SHUCKBURGH

Introduction

It is interesting to contemplate an entangled bank, clothed with many
plants of many kinds, with birds singing on the bushes, with various
insects flitting about, and with worms crawling through the damp earth,
and to reflect that these elaborately constructed forms, so different from
each other, and dependent on each other in so complex a manner, have all
been produced by laws acting around us ... Thus, from the war of nature,
from famine and death, the most exalted object which we are capable of
conceiving, namely, the production of the higher animals, directly follows.

(From the final paragraph to *On the Origin of Species by Means
of Natural Selection*, 1859.)

'Survival of the fittest' is probably the best-known reduction of Charles
Darwin's thought. The phrase was coined by the British economist Herbert
Spencer in an 1864 work, after reading Darwin's thesis, but it was adopted
by Darwin himself in later editions of his work, who said he found it 'more
accurate, and sometimes equally convenient' (5th edition, 1869).

Spencer used the phrase beyond the realms of naturalists to lend support
to his social theories. In *The Man versus the State*, 1884, he wrote:

And yet, strange to say, now that this truth [the survival of the fittest] is
recognised by most cultivated people, now more than ever before in the
history of the world, are they doing all they can to further survival of
the unfittest!

Survival, edited by Emily Shuckburgh. Published by Cambridge University Press.
© Darwin College 2008.

He goes on to explain this statement by saying that those who try to assist the 'unworthy' make the 'struggle for existence harder for the worthy' by inflicting on them 'artificial evils in addition to the natural evils they have to bear'. Thomas Huxley, a friend of Spencer and an ardent supporter of Darwin's theory of evolution, argued conversely that social organisation securing 'a fair amount of physical and moral welfare' was needed to mitigate against the natural, gladiatorial struggle for existence.

Others followed. Almost everywhere in Western civilisation thinkers of the Darwinian era seized upon the new theory and attempted to sound its meaning for their own disciplines. Life is struggle; and in that struggle the fittest survive: this concept pervaded all aspects of *life*. To the British imperial propagandists, a number of whom founded the National Efficiency Movement at the turn of the twentieth century (see G. R. Searle's book of that title), protecting culture, language, health, were component parts of the grand struggle for survival in the world at large. In America, captains of industry like John D. Rockefeller and Andrew Carnegie used it to justify laissez-faire capitalism. Thus Darwin's thoughts on survival were invoked *broadly*, not just in the context of understanding the natural world, but also in a host of other areas.

This book considers the 'survival of the human race' in the same broad sense as these Darwinist thinkers. By detailed examination of a number of very distinct aspects of human life, we will explore what strategies can be adopted to assist us in our individual struggle for existence and to preserve and indeed improve our collective lifestyles. The topic is vast in its coverage, and we will be restricted to examination of but a few distinct fragments.

We will start by examining the survival of one of the main types of polity within which humans have chosen or been compelled to live: empires (*Survival of empires*, Paul Kennedy). Expressions of identity and cultural heritage have been integral parts – and problems – not just of empires but of the entire human experience, and will form the subjects of the following two chapters (*Survival of culture*, Edith Hall and *Survival of language*, Peter Austin). Subsequent chapters (*Surviving disease*, Richard Feachem and Oliver Sabot, *Surviving natural disasters*, James Jackson and *Surviving famine*, Andrew Prentice) will consider questions of survival in the face of horrors that are both prosaic and profound for many – perhaps most – human beings, now as in the past. The final two chapters (*Surviving longer*, Cynthia Kenyon and Claire Cockcroft; and *Surviving into the future*, Diana Liverman) will conclude our journey by examining aspects of survival which have a distinctively modern feel: the biological challenge of living longer, and the future survival of societies on a planet influenced by climate change.

Threats to survival

Quite clearly we humans are to large extent the creators of our own destiny and the makers of our own doom. Natural selection there may be, but more often than not the selective pressure is now of an underlying anthropogenic cause. This allows for a gruesome and potentially catastrophic feedback, as poignantly described in a recent book entitled *Hegemony or Survival*, by Noam Chomsky. He suggests we 'are entering a period of human history that may provide an answer to the question of whether it is better to be smart than stupid', and suggests that if this question receives a definite answer it can only be that 'humans were a kind of "biological error," using their allotted 100,000 years to destroy themselves and, in the process, much else'.

Popular belief has us either invincible or en route to defeat. To some, the greatest woes of global society are so unimaginably vast that there isn't much we as individuals or even as collective bodies can do to ameliorate them; indeed much is utterly beyond our control. Others believe in the limitless ability of modern civilisation, with its scientific and technological capabilities, understanding of economic principles and knowledge gained from historical precedents, to overcome any crisis. Either one of these beliefs may ultimately prove accurate, but as we shall see in later chapters of this book, the problems of global pandemics, natural disasters and global climate change whilst being serious threats to our survival, need not be considered insurmountable. As Amartya Sen put it in *Development as Freedom*, 'Tacit pessimism often dominates international reactions to [the] miseries in the world today.' But, he goes on, 'there is little factual basis for such pessimism, nor are there any cogent grounds for assuming the immutability of hunger and deprivation'.

It is perhaps futile to try to tease out absolute causes of the many varied threats to survival, but some factors seem to be frequently present. One such factor is poverty, taken in its broadest sense to describe a deprivation of basic capabilities reflected in premature mortality, significant under-nourishment (especially of children), persistent morbidity, widespread illiteracy and other failures. Time after time we will see in the follow-ing chapters that it is the most poverty-stricken who are most vulnerable. In the final chapter, Diana Liverman talks about the 'double exposure' of vulnerable groups to the risks of climate change and economic instability. For these groups, poverty is inextricably linked to deprivation of economic and political strength, which in turn is linked to poor health and education and this is then linked back to poverty and reduced freedoms. Feeding off this depressing loop is a greater vulnerability to a broad range of threats to survival, impacting cultural identity as well as longevity. The effects of this cruel web of feedbacks are felt particularly in African countries ravished by the terror of HIV/AIDS as is potently described by Richard Feachem and Oliver Sabot in their chapter on *Surviving disease*. Indeed the millennium year opened with the United Nations Secretary-General Kofi Annan declaring the impact of AIDS in Africa to be 'no less destructive than that of warfare itself'.

Subjugation, be it in the form of internal repression by governing powers or of restrictive policies of colonial rulers, is another factor frequently asso-ciated with threats to survival to both the life of individuals and the culture of communities; Peter Austin describes the loss of language in this context in his chapter. In her chapter on *Survival of culture*, Edith Hall uses one of the most famous tales of subjugation – Odysseus and the Cyclops – to explore the interleaving between colonial oppression and cultural heritage.

In considering the collapse or survival of past societies, Jared Diamond in his books *Guns, Germs and Steel* and *Collapse* has emphasised the role of environmental factors often play. He puts forward five factors often contributing to the failure of societies: environmental damage, climate change, hostile neighbours, decreased support from trade partners and the society's response to its environmental problems. Indeed danger lies not only in current responses to environmental problems. In his chapter on *Surviving natural disasters*, James Jackson discusses the dangers posed by historical reaction of a society to their environment, exemplified by the engineers of ancient Persian civilisations, whose ingenious technology to bring water supplies to the deserts have resulted in vastly populated cities such as Tehran evolving in near-certain earthquake disaster zones.

Thus poverty, subjugation and environmental concerns are all factors associated with threats to survival, be they threats to the lives of individuals or to the cultural integrity of communities. If there is a virtue in finding a common linkage between all these various factors, then perhaps the best description follows Amartya Sen's thinking and considers the *restriction of individual freedoms*, taken to involve both the processes that allow freedom of actions and decisions and the actual opportunities that people have given their personal and social circumstances. He has famously commented that 'no famine has ever taken place in the history of the world in a functioning democracy', pointing out that 'authoritarian rulers . . . lack the incentives to take preventative measures' whereas 'democratic governments . . . have to win elections and face public criticisms, and have strong incentives to undertake measures to avert famines and other catastrophes'.

Yet in this soup of interdependencies, disasters themselves often bring about loss of law and order, and the breakdown of democratic stability. Andrew Prentice in his chapter on *Surviving famine* notes that in surveys carried out after famines, nearly everyone will admit that hunger drove them to theft. Similarly, the terrible plague of Athens in 430–427 BC resulted in a general lawlessness as is described by the Greek historian Thucydides in *The History of the Peloponnesian War*:

> No fear of god or law of man had a restraining influence . . . no one expected to live long enough to be brought to trial and punished: instead everyone felt that already a far heavier sentence had been passed on him and was hanging over him, and that before the time for its execution arrived it was only natural to get some pleasure out of life.

Shockingly, in the days following the 2005 hurricane in New Orleans the ensuing lawlessness was evident for the world to see with reports of shootings, carjackings and thefts across the city (see Figure 1.1). An article in the *Washington Post* entitled *A City of Despair and Lawlessness* started:

FIGURE 1.1 New Orleans, United States: People waiting to be evacuated from the Superdome take cover after the National Guard reported shots being fired outside the arena on 3 September, 2005, six days after hurricane Katrina hit the city. Some 3,000 people are still believed to be outside the Superdome. (Photo credit: Nicholas Kamm/AFP/Getty Images).

> NEW ORLEANS, Sept. 1 – Federal and local authorities struggled Thursday to regain control of this ruined and lawless city, where tens of thousands of desperate refugees remained stranded with little hope of rescue and rapidly diminishing supplies of food and drinking water.

There is one final factor of primary importance: population growth. In *An Essay on the Principle of Population* in 1798, Thomas Malthus made his dramatic and defining statement:

> The vices of mankind are active and able ministers of depopulation ... But should they fail in this war of extermination, sickly seasons, epidemics, pestilence, and plague, advance in terrific array, and sweep off their thousands and ten thousands. Should success be still incomplete, gigantic inevitable famine stalks in the rear, and with one mighty blow levels the population with the food of the world.

Although the Malthusian catastrophe has never materialised and population growth has failed to follow his exponential growth model, population size and growth cannot be ignored when considering threats to survival. On a crude and fundamental level, a larger population results in more people available to die as consequence of a threat to individual survival in one guise or another. This concept was captured by Pierre Verhulst in his logistic equation model of human population growth (1838). More subtly, the distribution trends of large populations influence the survival prospects. The inexorable move of peoples from rural to urban environments in the name of development is leading to what can only be described as mega-cities in many parts of the world today. In China urbanisation has been particularly remarkable – since the 1950s its urban population has increased nearly sevenfold to half a billion people, representing a third of the total population. Vast populations living in close proximity, often in environmentally ill-conceived locations. The increased risks are too numerous and too obvious to state.

Just one example comes from Amoy Gardens, a large housing estate in the Kowloon District of Hong Kong consisting of ten 35-storey buildings where around 15 000 people reside. During the SARS epidemic of 2003, a single person visiting the complex infected more than 300 residents in a matter of days, many from the same block, with the infection probably being spread through the plumbing system rather than person-to-person contact. Many observers link the tragic genocide that unfolded in Rwanda in 1994 to high population pressure. In the words of Gérard Prunier, 'The decision to kill was of course made by politicians, for political reasons. But at least part of the reason it was carried out so thoroughly by the ordinary rank-and-file peasants . . . was that there were too many people on too little land, and that with a reduction in their numbers, there would be more for the survivors.' (*The Rwandese Crisis, 1959–1994.*)

An interesting paradox of human nature is that we appear to be fascinated by the destructive power of the threats to the survival of others – one only needs to open a newspaper or switch on the television to find evidence of our obsession – whilst at the same time we appear to be somewhat oblivious of threats to our own personal or collective survival. This dichotomy sometimes manifests itself in 'them' and 'us' scenarios as exemplified by the initial attitude of many to HIV/AIDS: 'It's something that affects "them", not "us" (and it may even be their own fault).' Sometimes we simply are not aware of the threat that creeps upon us – the slow but

7

steady decline in numbers speaking a language. At other times it is as if we have an innate belief in our indestructibility – this feeling seems to be especially prominent within the present-day Western world. Freak occurrences destabilise us, but only temporarily – in recent years the terrorist attacks of September 11th 2001 and the New Orleans hurricane of 2005 rocked the American nation, and have had long-lasting implications, but the fear is fading and the feeling of collective invulnerability is returning.

Mixed with this is perhaps a notion of what might be labelled 'devolved responsibility' for addressing global problems, i.e. the notion that they are something the greater population, national governments or international organisations should concern themselves with, but mere individuals are in no position to take effective action over. After decades of support of capitalist ideals and the power of individual destiny, it is almost an ideological struggle for us to accept both individual and collective responsibility, social responsibility, for addressing global threats, such as climate change, which after all are rooted in the combined impact of individuals. But examination of past societies shows that collapse of societies, when it occurs, can be rapid, the most recent example being the seemingly near-overnight collapse of the USSR. Largely as a result of enormous social and economic dislocation, life expectancy declined dramatically in Russia during the 1990s – an unprecedented experience for an industrialised nation. Thus we should perhaps be a little more cautious.

Routes to survival

Arguably, the key human strategy to promote survival is to *organise*. We have been developing organisational skills since the introduction of agriculture some ten thousand years ago in the so-called Fertile Crescent. A quick review of our subsequent evolution is enlightening for it reveals this and other strategies for survival. The two civilisations that emerged in the fourth millennium BC – the Sumerians of Mesopotamia and the Egyptians – invented various techniques to produce more plentiful harvests, most notably irrigation. They also instigated long-distance trade to access key resources that were lacking in the region, and consequently communication and co-ordination became essential. This prompted an innovation of immense importance – writing – and with it the evolution of an organised social structure involving a complex hierarchy including professional scribes. Writing dramatically extended the collective memory of societies, enabling the transmission of knowledge useful for individual survival and the survival of a community's cultural heritage. It would also come to enable societies to retain closer cultural links with their, sometimes more glorious, past (the Egyptians of the seventh and sixth centuries BC harked back to the art and architecture of the third millennium BC, just as today we remain attracted to fashions of classical times). Social complexity and political organisation continued to increase such that the Sumerians were divided into some thirty city–states in the third millennium BC, and by the first millennium BC these had given way to empires like those of the Assyrians and Babylonians. Successful groups were those able to organise themselves to control and distribute resources. From there, political struggle, expansion and colonisation, education and democratisation resulted in the world as we know it today.

We can easily pick out from above the strategies our predecessors have used to be successful: organisation, communication and innovation. These are very human strategies, seen rarely elsewhere in the animal kingdom. Indeed there is some suggestion, discussed further by Cynthia Kenyon in her chapter *Surviving longer*, that the length of human lifespan – in particular our post-reproductive lifespan – is linked to this strategy. Moreover, we will see in future chapters that these strategies are essential in promoting survival in all its senses. The success of empires can be found in their logistics and communications networks, as Paul Kennedy describes in his chapter *Survival of empires*. Policies for promoting survival against disease, famine, natural disasters and climate change all must have these three strategies at their heart, so too policies for promoting the survival of language and other components of our cultural identity.

A note of caution though is necessary. In his book *The Collapse of Complex Societies*, Joseph Tainter suggests that such societies 'by their very nature tend to experience cumulative organisational problems'. By way of example he goes on to describe how 'as regulations are issued and taxes established, lobbyists seek loopholes and regulators strive to close these', and that with an increased need for specialists to deal with such matters, 'an unending spiral unfolds of loophole discovery and closure, with complexity and costs continuously increasing'. This example strikes a chord with the description of the political situation surrounding carbon emissions reduction mechanisms described by Diana Liverman in her chapter. Tainter argues that the return on an investment in complexity varies, and that this variation follows a characteristic curve where a situation of declining marginal returns is inevitably reached. At that stage he says collapse can occur from one or both of two reasons: 'lack of sufficient reserves with which to meet stress surges, and alienation of the over-taxed support population'. Thus too much organisation may generate its own threat to survival through social exhaustion.

Now let us consider some of the current threats to our survival. The population of the world presently stands at some 6.5 billion. That is an estimated 5.5 billion more people living on the Earth now than in 1800, and indeed a doubling of the global population since the 1960s. China and India lead the list of the most populous countries and are both growing fast. We are packing the planet ever more densely. Moreover as technology develops, we are travelling further and faster, mixing and integrating more and more. The result is that where once local problems were the most urgent for individuals, over the past decades, global problems have started to take precedence. Naturally concerns are raised as to the consequences of China, with its large population, succeeding in its goal of achieving First World living standards, and with them the First World's per-capita environmental impact.

How should we react to these global problems? If organisation is largely the key to our success then global problems should surely be addressed by global organisation. But if so, what form of global organisation? – the United Nations for all its worthiness hasn't exactly earned itself gold-star status as a global problem-solver. Indeed, increasingly it is non-nation–state actors who are seen to be addressing global problems: the Bill and Melinda Gates Foundation, now reputed to be the world's largest charity, paid out $1.36 billion in grants for a wide variety of causes in 2005 alone; the Global Fund, whose first executive director was Richard Feachem, has been set up as an independent body to spearhead the fight against AIDS, tuberculosis and malaria; a host of other charities and non-governmental organisations battle daily against threats to survival; and millions of individuals throughout the world have become involved in a variety of schemes, perhaps the most visible being the recent celebrity-backed campaigns to fight global poverty.

In the years following the Second World War, thinkers were understandably focused on the threat of future global wars and in particular the threat posed by nuclear weapons. Bertrand Russell argued that 'the survival of the human race depends on the abolition of war, and war can only be abolished by the establishment of a world government'. Albert Einstein too entered the debate with a newspaper article stating:

> ...a world authority and an eventual world state are not just *desirable* in the name of brotherhood, they are *necessary* for survival. In previous ages a nation's life and culture could be protected to some extent by the growth of armies in national competition. Today we must abandon competition and secure cooperation. This must be the central fact in all our considerations of international affairs; otherwise we face certain disaster. Past thinking and methods did not prevent world wars. Future thinking *must* prevent wars.
>
> ('The real problem is in the hearts of men', *New York Times*,
> 23 June 1946.)

He goes on to explain his opinion that although the danger (of nuclear weapons) had been brought forward by science, the real problem lay in the hearts and minds of men and that 'when we are clear in heart and mind – only then shall we find the courage to surmount the fear which haunts the world'.

The threat of nuclear war still lurks today in international debate as new nations develop, or try to develop a nuclear capability. Global law and order still remains at the top of the list of concerns for all governments, but we now also have a host of additional global problems to concern us. Some are universal problems and hence global in their distribution, others have arisen through the increasingly globally integrated world we live in, more still are simply inherently global problems – national boundaries are no barrier to the spread of atmospheric pollution or avian flu in migratory birds. Without global agreement it will be impossible to address these and other concerns. Global public goods are prone to market failures and can only effectively be addressed by the intervention of global government. And yet the experiences of those trying to obtain global agreement on global threats such as those concerning climate change or pandemics demonstrate that this is no easy matter. Richard Feachem and Oliver Sabot discuss in their chapter the issue how to determine priority in addressing such global threats.

One cannot help feeling that it should not be an either–or decision as to whether to address one global threat or another. And yet, financial resources and political will are finite. The notion of a global public good is a useful one, but there are many global public goods. In pitting pandemics against climate change against development goals, what metrics should be used to decide priorities? Mortality or potential mortality is an obvious choice. Richard Feachem and Oliver Sabot point out that already 25 million individuals have died from HIV/AIDS and some 40 million more are infected with the virus, and these numbers may be compared against those arising from certain other threats. But how can one quantify the mortality due to climate change? The heatwave in Europe in summer 2003 certainly resulted in an increase in deaths among vulnerable people – some estimate it at as many as 50 000. The summer temperatures were the highest in Europe for over 500 years and as Diana Liverman notes, studies have shown human-induced climate change to be implicated in increasing the risk of such a heatwave, but it is difficult to be definitive. Moreover, if we were to use 'lives saved' as our metric, what time-frame should we use for comparison? The next year? The next decade? The next century?

In trying to decide how to react to the new threat of global climate change we are forced into uncharted waters. The decision is particularly problematic for three rather unique reasons: thresholds, uncertainty and timing. Interactions within the Earth system are non-linear and hence thresholds can be reached beyond which there is no possible return to past states, and tipping points beyond which large-scale discontinuities can be expected. Such behaviour is highly non-intuitive to most. We are used to a world where the Sun rises every morning and sets every evening, to a world with only gradual and slight variation. Intuitively it seems very difficult to believe that, outside Hollywood, large-scale discontinuities could really start to ravage our planet. However, we are patently misguided in our intuition – large-scale discontinuities have demonstrably occurred in the geological past over short timescales. Future predictions of climate are highly uncertain making it hard to decide what the magnitude of mitigation or adaptation responses should be. Some of this uncertainty is due to our poor understanding or poor ability to model the science, and this slowly improves with time. Some of it is inherent and due to the chaotic nature of the earth system: the effect famously captured by the question 'Does the flap of a butterfly's wing in Brazil set off a tornado in Texas?' Thus we cannot escape the uncertainty, we are restricted, obliged to make political decisions taking into account the fact they will inevitably be based on uncertain information. The timing problem is one of the simple scientific fact of a time delay between kicking Mother Earth and hearing her scream. Present-day emissions will continue to alter the climate into the next century and beyond. Politicians though are understandably reluctant to act to calm the planet before they can hear her injured wails – again it goes against basic intuition. But as Diana Liverman notes in her chapter, delaying emissions reductions by just ten years will produce a much more serious warming and will require later a larger cut in emissions than if we act now. The Kyoto Protocol is a start, but its measures do not provide the reductions scientists believe are required, leading James Lovelock to declare in his polemic *The Revenge of Gaia* that Kyoto was 'a mere act of appeasement to the polluters'.

Problems of timing are not limited to climate change. All politicians have only a limited time in power. This inevitably introduces an artificial timescale into the reaction to a threat to survival. Even in a democracy, if a threat is likely to materialise only outside a political term, what possible incentive does a politician have to act? Despite the fact that some political leaders talk of their legacy, for the most part, politicians are focused on short-term voter satisfaction. This problem was recognised by Plato who discussed the desirable qualities of political leaders in the *Republic*, writing that until 'political greatness and wisdom meet in one...cities will never have rest from their evils, – nor the human race'. Perhaps we do need an alternative style of leader to address the troubles of the human race; perhaps we need wiser guardians with long-term visions. But perhaps also Einstein had it right. Perhaps the real problem (and hence the solution) lies with the people, and only when we voters are clear in our hearts and minds about the need to address long-term threats will our politicians be able to find the courage to surmount the fears which haunt the world.

So how can we start to reach the hearts and minds of the people? Role models clearly have an important part to play. As mentioned earlier, threats to survival too frequently seem so vast to a mere individual that they appear insurmountable. Role models can light a path up the hill. Peter Austin describes in his chapter repeated uses of role models in promoting the survival of endangered indigenous languages. Edith Hall describes how different communities have adopted either Odysseus or Polyphemus as role models for their own behaviour (be it good or bad). Paul Kennedy describes how the Spanish Empire harked back to the Roman Empire. Diana Liverman puts forward the '40% house' project as a role model for how to reduce our personal consumption of fossil fuels. And Cynthia Kenyon suggests that the key reason why we have yet to use science and technology to extend our lifespans further is that we lack a suitable long-lived role model from the animal kingdom – few aspire to the life of a giant tortoise or similar.

What about the other strategies we have developed for survival – communication and innovation? Communication networks are important for disseminating information, on for example safer sexual practices, and can help save lives in an immediate sense. Knowledge communicated through cultural heritage or tradition can also be critical. The Chinese Confucians continue to revere their ancestors and set great store by their ancient teachings, and indeed throughout the world tales abound of communities whose survival is attributed to knowledge obtained through the legacy of their ancestors (for example, ancient stories from Javanese mythology are said to have saved the lives of many locals during the Krakatoa volcanic eruption of 1883 and ensuing tsunamis, by raising their awareness of the risk of inundation from the sea). Nevertheless, cultural heritage and tradition can at other times be detrimental to survival, leading to an inappropriate response to a new threat. It is said that the failure of the Vikings in Greenland in contrast to the Inuit, for example, can be partly attributed to the knowledge, cultural values and preferred lifestyles they brought with them based on generations of experience in Norway and Iceland. Jared Diamond notes in his book *Collapse* that 'the values to which people cling most stubbornly under inappropriate conditions are those values that were previously the source of their greatest triumph over adversity'. This observation can certainly be applied to some of the threats to survival discussed in subsequent chapters, perhaps most significantly to climate change.

A developed education is needed for long-term progress (including innovation), indeed it has long been recognised as essential to the success of nations. Back in 1851, Prince Albert wrote in a memorandum for the Commissioners of the Great Exhibition that he was concerned Britain would be left behind as France and Germany were 'continually economising and perfecting [industrial] production by the application of science' and he called for more attention to technical education. Some years later the British Prime Minister Disraeli crystallised this idea in a speech in 1874 where he famously proclaimed: 'Upon the education of the people of this country the fate of this country depends.' Amartya Sen has taken this further and has argued more broadly that 'development can be seen as expanding the real freedoms [which include education] that people enjoy'. Education gives more power to the people, it allows for the democratisation of decisions concerning responses to threats to survival. James Jackson argues in his chapter *Surviving natural disasters* that with education the public in the developing world would start to realise that total destruction from earthquakes is not inevitable and would start to demand their buildings conform to modern standards. This pits political and economic expediency against individual freedoms; the survival of the fittest against the survival of the unfittest.

The value of simplistic, overarching conclusions is of course limited. Nevertheless, it is perhaps possible to extract a few general comments from the broad range of threats to survival that are considered here. Firstly, it seems that organisation, communication and innovation are universally important components of strategies to address threats to survival. Secondly, new and emerging threats with global dimensions self-evidently require global responses. And thirdly, role models, education and a sensitive treatment of cultural tradition must all play a part. Finally, in debating survival, one surely has to take care not to adopt an overly academic approach and instead to maintain a sense of humanity. Joseph Tainter states in *The Collapse of Complex Societies* that 'under a situation of marginal returns, collapse may be the most appropriate response'. He notes that such collapsed societies have not failed to adapt, on the contrary, 'in an economic sense they have adapted well', adding 'perhaps not as those who value civilisations would wish, but appropriately under the circumstances'. Which is all very well, but it fails to account for the human suffering that inevitably accompanies such a collapse.

Closing words

This book arose out of the *Twenty-First Darwin College Lecture Series* held in Cambridge in early 2006. As such, each chapter is an independent essay and readers may choose to treat them as such. But collectively they form a journey through human life with each chapter building on its predecessors and setting the scene for its successors. The interested reader will find a host of unexpected interconnections.

For the benefit of those wishing to pick and chose from the chapters, here is a brief synopsis of each chapter in the words of the authors.

Survival of empires, Paul Kennedy: 'The survival of the fittest' proved very compelling to politicians and publicists in the late-nineteenth/early-twentieth-century age of imperialism. Struggle was natural, and everywhere. Countries were either rising or falling. There was no standing still in the era of the 'Scramble for Africa', the Spanish–American War (1898), the Russo-Japanese War (1904–05), and the unprecedentedly bloody First World War. But how, then, were empires to survive? The overwhelming answer seemed to be, through organisation – through the harnessing of all of the resources of metropolitan society, and not just its military, naval, colonial, technological and financial resources, but much else besides. Many of the examples employed in the chapter relate to the policies for survival employed by the British Empire from Darwin's time to the end of the Second World War, but it also includes remarks upon other empires (Rome, Spain, the United States today).

Survival of culture, Edith Hall: The intimate link between the Renaissance and Enlightenment rediscovery of antiquity and the era of colonisation and empire has implicated the study of Greek and Roman authors in the legitimisation of the Western domination of the globe. Anti-colonial and postcolonial writers have associated the canonical European classics with both historical exploitation and ongoing threats to non-Western identity and indigenous culture. The transhistorical popularity and influence of the Homeric *Odyssey* has resulted in part from its glorification of the intelligent, culturally sophisticated travelling warrior who sacks cities and accumulates capital as he takes control of less developed distant shores. However, the Cyclops, a victimised and vilified figure, can perhaps also offer the postcolonial world a way of thinking about the survival of the 'master texts' of classical culture that is less threatening and alienating to those working for cultural survival in its other, more urgent sense.

Survival of language, Peter Austin: Across the world minority languages are under threat from larger regional and global languages as communities shift their preferences in favour of what they perceive as economically, politically and socially more powerful tongues. In the process languages become endangered as children are no longer learning them – eventually such threatened languages can and do disappear. This chapter addresses a number of issues: What are the factors that determine a language's survival? Are all smaller languages doomed to replacement by a few larger stronger ones? If a language is endangered is there anything that can be done to ensure that it does survive and does not become extinct?

Surviving disease, Richard Feachem and Oliver Sabot: All nations of the world now face a major challenge from global pandemics for which they are unprepared. The largest global pandemic in the history of humankind, the HIV/AIDS pandemic, continues to spread and to devastate. Meanwhile, the world faces the prospect of a huge and rapidly spreading pandemic of avian flu. Facing up to these and other pandemics is a global public good of the highest priority. Global capacities to deal with current pandemics, and to avert future pandemics, are woefully inadequate. The nature of determinants of global pandemics, past, present and future, are discussed and some of the lessons learnt from HIV/AIDS in relation to preparedness for avian flu are elaborated. The chapter includes comments on the need for enhanced supranational mechanisms to deal with existing and future global pandemics.

Surviving natural disasters, James Jackson: In the last few decades several devastating earthquakes have apparently targeted population centres in otherwise sparsely inhabited regions, particularly in Asia.

Ancient settlements are often located for reasons to do with water supply, access, strategic defence or control of positions on trade routes, and these considerations are, in turn often determined by natural geological phenomena, particularly features of the landscape created by earthquakes. It is this close relation between where people live and earthquakes that leads to the apparent bull's-eye targeting of cities by earthquakes. As a result, we should expect many more disasters this century, some of which will be far worse, in terms of mortality, than those we have already seen. The question of what to do with the huge populations concentrated in earthquake-prone mega-cities of the developing world has no easy solution.

Surviving famine, Andrew Prentice: Who amongst us has really been forced to contemplate the dark horror of famine? Yet to our recent ancestors the unimaginable horrors of famine have been an ever-present threat. Somewhat paradoxically, it was the dawn of agriculture that heralded seasonal hunger and catastrophic famines caused by climatic instability. Huge populations dependent on a single staple crop could be devastated by drought or blight. And if they escaped the wrath of nature they could be scythed down by man's inhumanity to man when starvation was used as an instrument of war and subjugation. Surviving famine has driven the evolution of a range of metabolic and behavioural adaptations. In modern wealthy societies these responses are redundant and many have become maladaptive, driving the pandemic of obesity and diabetes. We bear the mark of our ancestors' struggles against famine indelibly etched into our genome.

Surviving longer, Cynthia Kenyon and Claire Cockcroft: Ageing has long been assumed to be a passive consequence of molecular wear and tear, counteracted by the force of natural selection. But it's not so simple. The ageing process is under exquisite regulation by a complex, multifaceted hormonal system. By manipulating genes and cells in roundworms, we have been able to extend the lifespan and period of youthfulness of the worms by six times. We have found that signals from the reproductive system and sensory neurons influence lifespan by coordinating, via the hormone system, the expression of a wide variety of subordinate genes. Some of these subordinate genes can also influence the rate of onset of age-related disease. In this way, this hormone system couples the natural ageing process to age-related disease susceptibility. This work may lead one day to novel therapeutic or lifestyle approaches that allow us to stay young longer.

Surviving into the future, Diana Liverman: Climate change may present one of the greatest challenges for the survival of people and ecosystems in and beyond this century. The climate is already changing and the changes may be rapid and irreversible. Research on the impacts of climate change has expanded to assessments that include human health and ecosystems, and that focus on how vulnerability and resilience are the key determinants of damages. Current responses are clearly inadequate to the magnitude of the threat. This chapter argues that avoiding dangerous climate change requires a much greater effort that includes 60 % or more reductions in emissions, policies that include regulatory as well as market mechanisms and controversial technological decisions, and major commitment to planning and funding adaptation and to meeting the costs of damages through insurance or litigation.

I'll end this introductory chapter as it began, with the words of Charles Darwin for, as will become evident, his spirit drifts through every page of this book.

> Man may be excused for feeling some pride at having risen, though not through his own exertions, to the very summit of the organic scale; and the fact of his having thus risen, instead of having been aboriginally placed there, may give him hopes for a still higher destiny in the distant future ... But we must acknowledge ... that Man with all his noble qualities, with sympathy which feels for the most debased, with benevolence which extends not only to other men but to the humblest living creature, with his god-like intellect which has penetrated into the movements and constitution of the solar system – with all these exalted powers – Man still bears in his bodily frame the indelible stamp of his lowly origin.

Let's hope that despite our lowly origins, we use our acquired exalted powers, with continued sensitivity to the humblest living creature, to overcome the threats to our survival and win the struggle for existence.

FURTHER READING

Cook, M. A. (2004). *A Brief History of the Human Race*. London: Granta.

Diamond, J. (1997). *Guns, Germs and Steel: The Fates of Human Societies*. New York: W. W. Norton.

Diamond, J. (2005). *Collapse: How Societies Choose to Fail or Survive*. London: Allen Lane.

Tainter, J. (1987). *The Collapse of Complex Societies*. Cambridge: Cambridge University Press.

Sen, A. K. (1999). *Development as Freedom*. Oxford: Oxford University Press.

Sen, A. K. (1981). *Poverty and Famines: An Essay on Entitlement and Deprivation*. Oxford: Clarendon Press.

2 Survival of empires

PAUL KENNEDY

Introduction

The theme of this book, not to say the topics of the individual chapters, reflects the sheer range and interdisciplinarity of Charles Darwin himself and, perhaps even more so, the multitude of ways in which Darwinian thought and Darwinian expressions have been used, and misused, over the past 150 years.

In this chapter I will discuss 'survival of empires'. The intention is to look at the political and ideological movements that existed in Darwin's times, that were themselves evolving because of demographic, social, economic and technological change, and that found great utility in borrowing from and sloganising some of Darwin's best-known phrases, in particular, 'the survival of the fittest' and 'the struggle for survival'. It will discuss those campaigns for reform and renewal, in other words, for adaptation, since those political movements very much invoked what they regarded to be the 'lessons' of Darwin. It will focus in the first instance upon the case of Great Britain, because the political discussions in that country over the implications of Darwin's writings were to be intimately linked with almost a century of debates about and, more importantly, policies regarding, the survivability of the British Empire itself. But, in order to give a comparative dimension to such an enquiry in the manner of which Darwin would have approved, this article will also offer commentaries upon the survivability of various other empires, especially the Roman, the Spanish and the contemporary American.

Survival, edited by Emily Shuckburgh. Published by Cambridge University Press. © Darwin College 2008.

Historically and politically, there could be and indeed were two very different, actually, very contrary lessons to be drawn from Darwin's scholarly writings. These might, roughly speaking, be termed the passive and the active responses. In many ways, the passive one seems more natural. After all, if one reflected on 'the blind processes of natural selection', to use Professor Geoffrey Searle's term, and if one considered slow changes in species that had taken millennia of mutation and chance and climatic alterations, then surely the conclusion to be drawn was that there was nothing that could be done by human agency. These were changes on so vast and so undetectable a scale that, even if the science-passionate Victorians were just beginning to understand parts of that story, they were also coming to see that deep geological processes were beyond human control. Then there was the randomness. If it was true that a giant meteor did indeed crash into the Earth and kill off all the dinosaurs, then what political policy, favouring either the Liberals or the Conservatives, could be advocated on the basis of that fact? Darwin's reports that newer species of finches, or queer amphibian creatures, existed on the Galapagos Islands were interesting to those interested in such things; but Western sailors had been bringing back queer-shaped fish and lizards and mammals for hundreds of years. No doubt there would be more to come. But they were just curiosities, for the zoo or the dissecting table. It didn't affect us. And it didn't affect the British Empire in any policy-relevant way.

It is not this passive interpretation of the idea of evolution that is the focus of the present paper; if it were, it would be of very short length. Rather, it is those active or, if you like, reactive interpretations of Darwin's writings that will be dealt with here, and precisely because they did indeed have political, social and strategic reverberations; because they provided reference points for the debates about the future of the British Empire, and of all empires; and because they cast enormous and influential shadows upon the unfolding twenty-first century.

Let us consider for a few moments the landscape of international and national politics during the second half of the nineteenth century, indeed, up and into the First World War itself, by which time ideas about 'the survival of the fittest' had firmly entered the popular language and were earnestly debated, not just among intellectuals but among parliamentarians, the military and the business world.

It was a time, not of, say, French Revolutionary turmoil, but certainly one of massive wrenching changes in Europe, the United States, Russia and Japan, of changes that were occurring so fast and in so many different aspects of life and society that it was impossible to understand them all. It was an age of flux, not the first nor the last in history, but that was no consolation to contemporaries. The issue to them was that their way of life was being transformed, nationally and internationally, and what should they be doing about it.

Since our theme is the survivability of empires, this chapter can be brief about aspects concerning the domestic scene in Britain, noting them simply to remind readers of the variety of changes and of the uncertainties which, collectively, they induced. There was the fourfold increase in the total population of the United Kingdom during the nineteenth century, not leading to the catastrophes Malthus had forecast, but nonetheless to severe social and environmental strains, especially in the turbulent, unhealthy cities. There was the spread of industrialisation, the steam engine, the railways. There was the rise of organised labour, a new radicalism, socialist ideas, the decline of landed society, the coming of the plutocrats, and the emergence of a popular press catering to the newly educated masses. There was a decline in religious faith. After all, if you accepted that humans were descended from various categories of apes that flourished millennia before Adam and Eve, then where exactly did that put the Garden of Eden story? Where did it put God? If you accepted that human beings were like animals, struggling in the jungle, each to survive, then where did that put Enlightenment rationalism or Utilitarian beliefs in progress? The change of mood between young Tennyson's 1837 poem *Locksley Hall* and his work including the same title a half-century later captures it all. One scarcely needed Darwin to add to one's increasing religious doubts or to confirm that life was a struggle. But contribute he did, and society was ready to seize upon and, most usually, mangle into populist form what the newspapers told it Darwin's scientific researches had 'proved'.

The international scene provided even more dramatic evidence that the story of humankind was one of struggling for survival. The industrial and technological and communications revolutions were altering the world's power balances faster than ever before. Victorian Britain's economy was still growing, but those of the United States and Prussia–Germany were growing faster, and even Russia and Japan were moving through their industrial 'take-off' stage, just as, at the same time, other nations were falling behind. Nation–states were egoistic, turf-conscious, predatory, like animals in the jungle. Struggle was natural. There was no standing still in the era of the Scramble for Africa, the defeat of Spain in the Spanish–American War of 1898, the triumph of Japan in the Russo-Japanese War of 1904–05. A people's fate was decided through war, and there were plenty of wars around to prove that point conclusively. The 'Social Darwinists', as they came to be called, relished all this. Here is one, the French intellectual J. Novicow, writing in 1886:

> Nature is a vast field of carnage. Between living creatures conflict takes place every second, every minute, without truth and without respite. It takes place first between separate individuals, then between collective organisms, tribe against tribe, state against state, nationality against nationality. No cessation is possible. (Quoted on p. 86 W. L. Langer (1965) *The Diplomacy of Imperialism 1890–1902*. New York: Alfred Knopf.)

Even the cautious Prime Minister and Foreign Secretary, Lord Salisbury, who despised the Social Darwinists for their belligerency and ignorance of science, allowed himself the observation in 1898, during the debates on the carve-up of China, that 'You may roughly divide the nations of the world as the living and the dying...the weak states are becoming weaker and the strong states are becoming stronger...the living nations will gradually encroach on the territory of the dying and the seeds and causes of conflict among civilised nations will speedily appear.'

To Salisbury's more alarmed critics, however, the Prime Minister seemed to be missing the main point. It was not the fate of the decaying Chinese Empire, important though that might be, that was the real concern for the Social Darwinists; it was, rather, the preservation, that is, the survivability, of the British Empire in an era when the productive balances were shifting, and swiftly, to America, Germany, Russia and other large powers. Could the island–nation keep its place? The great editor of the London paper *The Observer*, J. L. Garvin, captured their deep-seated fears well in his brilliant essay of 1905, when he began with the sentence:

> Will the Empire which is celebrating one centenary of [the Battle of] Trafalgar survive for the next?

Survive, indeed. It should come as no surprise that Garvin's lengthy examination of that question was itself entitled 'The maintenance of Empire'. The concern of this influential school of Social Darwinists goes, in other words, to the core of this comparative survey.

Well, how could Britain and her Empire survive? These thinkers were not pessimists or, to use another term, 'declinists'. They eschewed passivity and inevitabilism. On the contrary, they advocated a whole host of positive policies, policies which, they held, would work provided everyone would agree to pursue them. Only a few of those policies can be listed here but the key to everything for them was to *organise* everything. Empires could only be made to survive, generation after generation, through superior organisation. The answer to Garvin's critical question – which of course he also sought to argue in his essay – was that the Empire could indeed maintain itself, even flourish, provided it was much better organised; provided it cast aside older, mid-Victorian habits and attitudes and prepared for the real, harsh world of the twentieth century. It all depended upon the British nation itself.

What was required, then, was the harnessing of all of the resources of metropolitan society and of the British-owned lands beyond, and not just the military, naval, colonial, industrial, technical and financial resources, but much else besides. Because Britain and its Empire were organisms, every part of the body had to be kept strong and vital. The educational system had to be thoroughly shaken up, from the infant schools to the ancient universities. New sciences had to be encouraged. Imperial College was founded. Latin and Greek had to give way to chemistry and electrical engineering, for how many battleships could a Department of Classics build? The Darwinists' criticism was unrelenting, very like the criticisms about British education by Mrs Thatcher in the 1980s, which was also unsurprising because she had been influenced by ideas from this Edwardian 'national efficiency' school that percolated through minor channels during the first half of the twentieth century, and re-emerged thirty years ago in Correlli Barnett's book *The Collapse of British Power* and in Martin Wiener's study of *English Culture and The Decline of the Industrial Spirit*.

Their Poet Laureate was Kipling, ridiculing the 'flannelled fools before the wicket' and rugby-playing 'muddied oafs', but praising the engineers, the railway-builders of India, the naval crews in the bleak North Sea.

Public health was also a great issue for the British Social Darwinists; high infant-mortality rates, stunted youths, men worn out by their thirties, on such weak foundations an Empire could not be sustained. The great cultural heritage of the island state, the language of Shakespeare and Gibbon, the importance of history – British constitutional and imperial history – had to be reaffirmed. Patriotism was a good thing. It bonded the nation. The future of Britain's youth was a special concern. How could they be trained to understand all this, to keep fit for the challenges ahead, to stay strong for the Empire? Baden-Powell's Boy Scouts Organisation was not, in its deeper purpose, a weekend outdoor-relief movement for inner-city youths, dedicated to spotting different trees and tying rope-knots. It had far higher aims in mind, nothing less than the long-term survival of the Empire. All this hung together.

If one failed to do all this, if one failed to put these reforms together, then one's days as a great empire were numbered, because there were always newer, rising empires, some of which gave signs of organising themselves better than you were – the Americans in their industrial productivity, the Prussians in their military power and then their rising High Seas Fleet, just across the North Sea; the Russians in their steady advance towards India. All these challenges had to be met, and roughly at the same time.

This theme of the 'survival of empires', in that it focuses (so far) upon the British debate of a century ago, may well lend thought to at least some of the chapters which are presented later in this book. The issues of good health or poor health, the critical matter of natural resources, the survival of culture, and of the language (or various languages), would have been inherently important to those Darwinist circles – and not just in Britain, but in Germany, America, Japan and other organised nations that sought to expand and maintain their empires. And who sought to ensure, above all else, survival.

It is now appropriate to consider the various manifestations of that term which so concerned the British imperialists – organisation – and to study this through the use of some pertinent historical examples. The four case studies selected here, Rome, Spain, Britain and the United States today, all offer us superb data about how their empires functioned, and how they worked as an integrated whole. They offer us much data about their communications and intelligence and logistics systems; they show to us, to employ another term, the nervous systems without which great empires cannot survive. What follows therefore has little or nothing to do with the ideology and rhetoric of empire, or about training fresh generations to run the empire, or about the cultures and literature of empire. All of them have their own mass of writings and studies. This contribution is more narrow; it is about how great empires organised themselves to keep on going, generation after generation, century after century.

Rome

Rome offered the best-known example here, and not just for Garvin's generation but for so many other national leaderships who came under stress to preserve their Empire, and looked to the past for guidance. How on earth did it last so long, and rule so successfully?

One answer, among many, must be that the *builders* of the Roman Empire were so good. Those hard-working Roman infantrymen – not slaves or hired hands, but the soldiers themselves, depicted here on Trajan's Column (Figure 2.1) – ensured that roads, bridges, aqueducts, fortifications were made to last for a long time into the future. What was the point of having local levees assemble, say, a bridge over the Rhine that might

FIGURE 2.1   Trajan's Column.

be swept away in the spring floods? Why not get the job done well in the first place, especially since the soldiers themselves – or their sons and grandsons – might be manning those very bridges and ramparts fifty years later? There is a method in all this that invokes admiration some two thousand years later. Even the Roman military camps, established everywhere from Hadrian's Wall to Carnuntum to North Africa, were of a standard design, as Polybius explained, so that a newly arrived unit, coming to its base perhaps as night fell, knew exactly where to go.

These bases, and the cities which often grew out of them, and the roads, bridges and harbours that linked them, were so strong and resilient that they would be used for many centuries after Rome fell by successor peoples whose own organisational capacities for creating such a network were negligible. Each piece of this 'nervous system' was linked to the other, and offered reinforcement to the other, in a way that made the whole so much stronger even than its many imposing parts. The after-dinner board game 'S.P.Q.R.' (Figure 2.2) is in an obvious sense just that – a game. But the

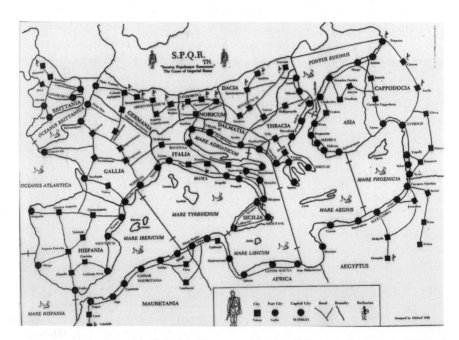

FIGURE 2.2   The after-dinner board game 'S.P.Q.R.' (Designed by Michael Mills; © Suffern, NY, 1982.)

layout of the board is actually very historical and the premise of the play is true to reality: you could move your players (i.e. legions) from one point to the next in a very short time.

Indeed, the board game's basic hypothesis is confirmed by a examination of the maps easily available to us in historical atlases and specialist studies of Roman grand strategy. The depiction of the Roman Empire at its greatest extent, around AD 117, is very well known and still captures our awe (Figure 2.3). But what is arguably a more interesting exercise is to superimpose upon that map a chart of the disposition of the twenty-seven Roman legions just a generation earlier, around AD 80 (see Figure 2.4). All the existing cities, bases, roads and shipping lines are still there, of course, and the general economy and the water-borne trades are flourishing, though none of those activities are visible on this map. But such activities rarely needed protection, and so the Roman garrisons are not stationed in those peaceful regions. They are, instead, deployed along those frontiers where the enemies of Rome pose significant threats – Britain,

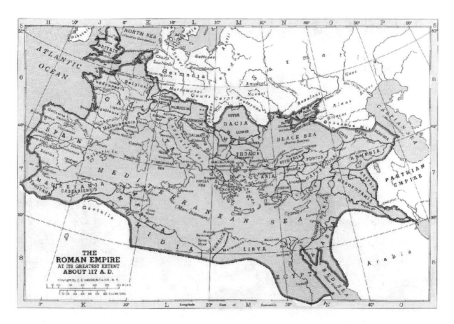

FIGURE 2.3 The Roman Empire at its height, AD 117. (*Hammond Historical Atlas of the World*, 1987.)

FIGURE 2.4 Disposition of the legions, *c.* AD 80. (Modified from *Atlas Zur Weltgeschichte*, p. 94, volume 1 (1964). Munich.)

all the way up the Rhine, the Danubian/Dacian frontier, and those Levantine borders beyond which Persia and other foes have their own powerful armies. The disposition of these legions, especially those along the Rhine, suggests that an attack on any one of them by the Germanic tribes could bring reinforcements within less than a week, perhaps within mere days, especially when one recalls that this communications network was equipped with an imperial messenger service (with regular stabling stations) to enable reports and orders to be transmitted with great speed. Cynics will also observe that none of these legions are, as yet, stationed near Rome or even in Italy itself; most governments, whether under the Republic or the Empire, worried that ambitious generals at the head of a legion or two might be tempted to meddle in domestic politics.

This was, then, a system with extraordinary organisational strength and thus, durability. Thus, the largest question, posed by Gibbon but also of course by the later Romans themselves, is, could it have endured even longer than it did? Or was it simply impossible, physically, to

impose rule inside so many lengthy frontiers and vis-à-vis so many dis-
parate enemies for ever and ever? The enormity of the crisis in the
later third century, and especially the sheer number of barbarian attacks
around a single year, AD 260 (Figure 2.5), was a forerunner of those
later waves of assaults that eventually caused the Empire to buckle
at the centre and to surrender province after province, leaving only
the Eastern Roman Empire intact but Italy and all of Western Europe
overwhelmed.

Was this simply a case of 'imperial overstretch' or, in another term, of
'the laws of physics', so that even Rome's sophisticated structures could
not maintain its grandiose world against the increasing pressures from
without? Or was there a military explanation for the steady weakening of
this great experiment in empire-building? On the latter issue the experts
differ greatly: Was the so-called 'grand strategy of The Roman Empire'
undermined by changes in the disposition of the troop units, by bringing
legions back from the frontier, by the 'barbarising' of the Roman Army
or the increasingly sophisticated 'Romanisation' of the barbarian forces
themselves, or by failing to keep a large mobile reserve that could be
thrown decisively at a threatened front? One doubts if any single explanation can be sufficient: military mistakes and weaknesses, rising economic
costs, divisions among the leadership, the growing fissures between the
eastern and western empires, possibly Christianity itself, are all part
of this epic story. The real achievement pertains not to the question
of how Rome sustained itself, but of how it survived, in vital form, for
so long.

Spain

This was a question that attracted students of politics in sixteenth-century
Europe, especially as they contemplated the rise and rise of Imperial
Spain. To be sure, the whole age of the Renaissance was fascinated by the
Greek and Roman worlds, and sought to emulate so many aspects of them;
from the disciplines of philosophy and architecture to apparently mundane
texts on military drill, contemporaries felt that they had much to learn
from the classics. Until the early to mid sixteenth century, however, they
had never had to contemplate the prospect that one of the many warring

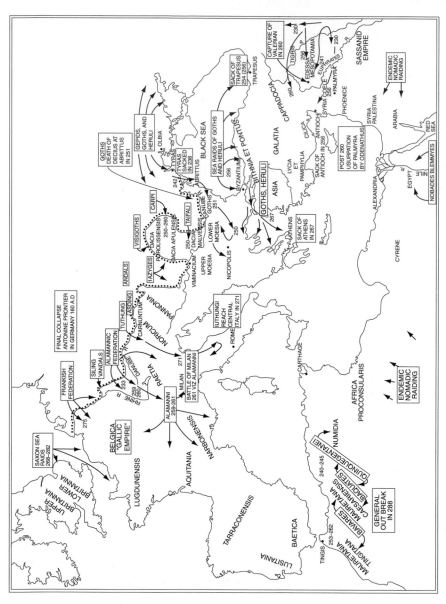

FIGURE 2.5 The Great Crisis of the third century AD. (Edward N. Luttwak (1976). *The Grand Strategy of the Roman Empire: From the First Century AD to the Third*. Baltimore: Johns Hopkins University Press, pp. 148–149.) © 1976 The Johns Hopkins University Press. Reprinted with permission of the Johns Hopkins University Press.

European states might come to dominate the continent. Now, the Spain of Philip II and his successors looked both ready and potentially capable of doing such a thing.

Explaining the emergence and relentless growth of the Spanish Empire is beyond the scope of this chapter. Instead we focus on its survival. What can be asserted, surely, is that this was a 'system', a trans-European polity that was held together by a network of dynastic territories, alliances, tax contributions and, as always and necessary, by sea-lanes and military routes across land. The most famous of the latter was the 'Spanish Road' (Figure 2.6), a logistical and diplomatic tour de force that allowed Madrid to project its power and influence into the Low Countries for close to two centuries. Spain's best and most reliable troops, raised and trained in Castile, would march to Barcelona, be transported by ship to Genoa, and then move north, either through the older Spanish roads in

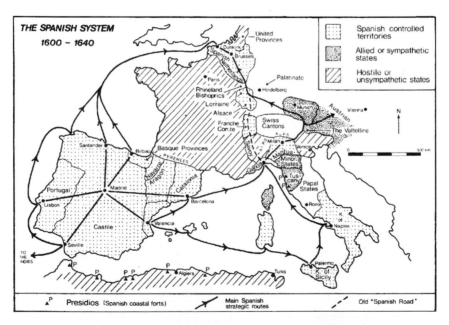

FIGURE 2.6   The Spanish Road. (R. A. Stradling (1981). *Europe and the Decline of Spain.* London: George Allen & Unwin, p. 54.) © 1981 George Allen & Unwin. Reproduced by permission of Taylor & Francis Books UK.

Savoy, Franche Comté and Lorraine, or via Milan and through the Swiss valleys to the friendly Austrian Hapsburg territories before proceeding down the Rhine. Flanked by various hostile or traitorous states, this was sometimes a precarious route; but the system generally functioned well, and was much more secure than sending troops via the Bay of Biscay and the Channel in the face of Dutch and English naval forces. At the same time, as the map shows, what we also have here is a nervous system that linked Spain – its troops, its envoys, its many trades – to its vital possessions in Italy, Sicily and Sardinia, gained from their revenues, and assured them protection from the great Ottoman danger in the east and south.

But this was not just a Europe-wide Empire. Indeed, Spain's greatest claim to fame, historically, was that it became the world's first truly global imperial system, whose ships literally went around the Earth (Figure 2.7). Already commanding a realm overseas that reached westwards to the Caribbean, Mexico and Peru and eastwards to the Philippines, Spain augmented her possessions by another great leap with the acquisition in 1580 of Portugal and thus of the latter's trading posts in Africa, India, the East Indies and Brazil. Here was a management challenge on the largest scale, one that might have daunted the ablest of Roman administrators. What is certain, at any rate, was that no regime in the history of the world had ever been presented with such an array of lands AND with the grand-strategical task of ensuring such an Empire's survival. In this respect, one can safely say that there were no historical precedents; Polybius, or even Thucydides, were limited in their utility here. Madrid was sailing

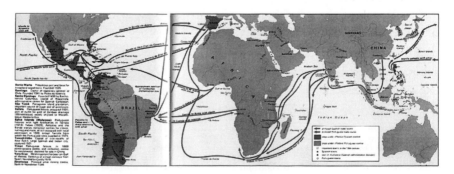

FIGURE 2.7 Networks in the Spanish Empire. (*The Times Atlas of World History.*)

into new waters, not just literally (e.g. in the North Pacific), but also in a larger, metaphorical sense: just how did one keep a worldwide empire going, generation after generation?

There were two, very contrary ways of looking at this new phenomenon. The first was to marvel at this agglomeration of power, either from the viewpoint of those pro-imperialist writers who rejoiced at the coming of a 'world monarchy', or from that of Spain's opponents, one of whom, the Elizabethan writer Francis Bacon, was openly distraught at this bid for global mastery and wrote (in 1595):

> France is turned upside down; Portugal usurped; The Low Countries warred upon; The poor Indians are brought from free men to be slaves. (See Paul Kennedy (1987). *The Rise and Fall of the Great Powers*. New York: Random House, p. 35.)

Yet there were others, especially high officials in the Spanish administration who had to grapple with the day-to-day running of this vast, uneven, over-extended machine, who were aware of its weaknesses and inefficiencies, and even more aware of the precariousness of the Empire's finances. This was an imperial system which, naturally, needed constant nourishment; and even the vast flow of income from Peru, the Philippines, the Italian estates, and Castile itself, could not keep pace with the income-devouring wars in the Mediterranean and Atlantic, against the Turks, in Germany and (especially) in the Eighty Years War against the Dutch rebels. Almost everyone, or so it seemed, wanted to pull Spain from its lofty pedestal.

And the blunt fact was that, while every one of Spain's foes could take a 'breather' sometime (even the Dutch; even the Turks) through an armed truce, Madrid could never order any significant reduction in her war effort because a new front had opened, or reopened somewhere else, just as the latest truce had been negotiated. The following chart speaks volumes (Figure 2.8).

Spain, then, unlike the other three empires examined in this survey, was never free from war – and by this we mean major war, not Rome's skirmishes in Wales or along the Danube, or Britain's post-1815 colonial campaigns in Africa. It was therefore unsurprising that from time to time the ministers of kings Philip II, III and IV (who reigned successively from

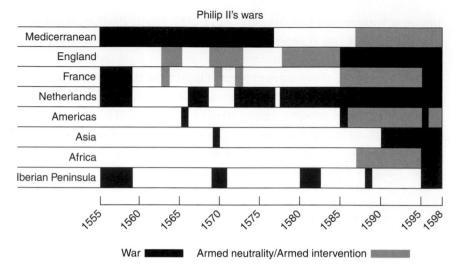

FIGURE 2.8 Philip II's wars. From Parker, G. The Grand Strategy of Philip II. (New Haven: Yale University Press, 2000) p. 2. © 2000 Yale University Press. Used with permission.

1556 to 1665) considered advising a permanent 'pull-out' from at least one of the main theatres of war. By so withdrawing, the argument went, the Empire could concentrate its resources on the truly vital tasks and thus ensure its survival, indeed, its victory. But the problem was that, precisely because the Spanish Empire was so like an organic system, it could not lop off one part of the body without affecting the others, perhaps fatefully. A now-famous internal memorandum of 1635 put this dilemma so well:

> The first and greatest dangers are those that threaten Lombardy, the Netherlands and Germany. A defeat in any one of these three is fatal for this Monarchy, so much so that if the defeat in those parts is a great one, the rest of the monarchy will collapse; for Germany will be followed by Italy and the Netherlands, and the Netherlands will be followed by America; and Lombardy will be followed by Naples and Sicily, without the possibility of being able to defend either. (See Paul Kennedy (1987). *The Rise and Fall of the Great Powers*. New York: Random House, p. 51.)

And so the Empire soldiered on, in so many ways doing amazingly well in its resistance to numerous foes (and of course still expanding in the Americas). As a nervous system, and a fighting system, its capacity to endure was remarkable. And, appropriately enough, when the end came it was not in any one theatre of war but in many, inside Spain and without. The shattering 'revolt of the Catalans' in 1640 coincided with the Portuguese decision to fight for independence, plus serious disturbances in Aragon and Andalusia, which then were followed by the slightly later uprisings in Naples and Sicily, the crushing of the Spanish (i.e. Castilian)

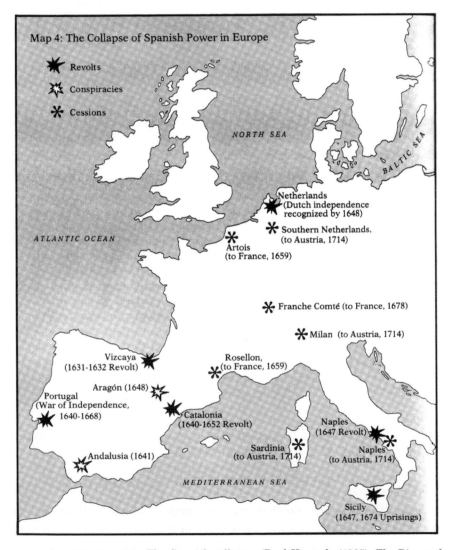

FIGURE 2.9 The Spanish collapse. (Paul Kennedy (1987). *The Rise and Fall of the Great Powers.* New York: Random House, p. 42.)

infantry at Rocroi in 1643, and the constant inroads of the French and the Dutch into Hapsburg domains (Figure 2.9). The recognition of Dutch independence in 1648 confirmed all this. The Eighty Years War had been a lost cause, a Vietnam disaster to the nth degree. This was not, of course, a total eclipse. World empires die more slowly, haphazardly, and in parts. Thus Spain would still be a great power for some time to come, but its place as Number One had evaporated.

Britain

Spain's fate in this two-century tale of great-power struggles was taken to heart by many a commentator in later times. Among those, ironically, was a British historian writing rather poignantly in the mid-eighteenth century about the decline of the Spanish Empire:

> That extensive monarchy is exhausted at heart, her resources lie at a great distance, and whatever power commands the sea, may command the wealth and commerce of Spain. The dominions from which she draws her resources, lying at an immense distance from the capital and from one another, make it more necessary for her than for any other State to temporise, until she can inspire with activity all parts of her enormous but disjointed empire. (See Paul Kennedy (1976). *The Rise and Fall of British Naval Mastery*. London: Allen Lane, p. 347.)

This was not of course a fate that the author assumed could ever befall his own island nation, which had just emerged from the Seven Years War (1756–63) with its own power more firmly established than ever over India, Canada and the West Indies. The British Empire was on the rise and, despite the setback (or blessing) of the loss of the American Colonies two decades later, the expansion overall would continue, generation after generation, and was still occurring in the year that Garvin offered his anxious question regarding the future survival of the British Empire. So here again we have a great case – the greatest of all in Western history, after the Roman – of the longevity and durability of an empire that stretched across many borders, indeed stretched across all seas and into every continent.

As with the other case studies, we will not discuss in this text the reasons for the rise of England, then Britain, then the British Empire; the literature, ranging back to Seeley's classic *The Expansion of England* and coming forward to the exhaustive new five-volume *Oxford History of the British Empire*, is prodigious and can satisfy any and all curiosities. Our focus is the Empire at its height and how it – like Rome and Spain – sustained its extensive imperial system for centuries. How was the Empire upon which 'the Sun never set' under-girded; how did its nervous system (its intelligence and communications networks) operate, and how well did they do? How did the centre reinforce distant points which came under threat? How, indeed, did it handle multiple threats?

The best clues come, as ever, from a study of imperial logistics and overseas bases and the links between each part of the whole. The red-tinted world map of the parts of the British Empire at the time of Queen Victoria's Diamond Jubilee of 1897, or of the Empire at its largest extent immediately after the First World War, are famous; even postage stamps copied this map (Figure 2.10). But any strategist would, like Garvin and

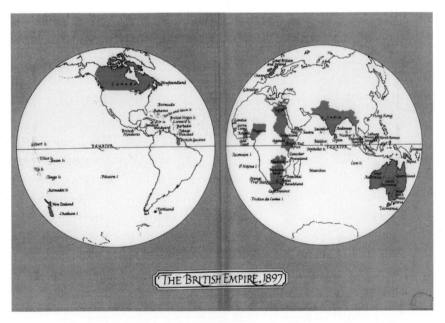

FIGURE 2.10 The British Empire in 1897. (J. Morris (1968). *Pax Britannica: The Climax of an Empire*. London: Faber & Faber.)

his colleagues, point out that such an array of imperial real-estate was akin to Spain's remaining empire in the 1760s, awfully scattered, potentially victim to local aggressors, and in most cases absolutely dependent upon control of the sea-lanes. Yet this tyranny of distance was what generations of British statesmen, admirals and generals had been planning to combat, through heavy and unremitting peacetime investments in naval bases and in 'all-red' undersea cable communications that would constitute the most modern equivalent of a strategic nervous system, bringing messages instantly from Whitehall to Capetown, Delhi and Sydney – and, of course, the reverse (Figure 2.11).

When these two maps are merged, then the British imperial 'system' virtually explains itself (Figure 2.12).

Examined as a whole, as was done with increasing frequency by the Committee of Imperial Defence after its founding in 1904, the integrated and mutually self-reinforcing nature of this organism becomes clear. At its centre, still, was the British industrial, shipbuilding, commercial and financial heartland of this otherwise decentralised worldwide body. To the many parts of the Empire flowed the home country's goods, services and emigrants, carried in British ships, insured by Lloyd's of London and

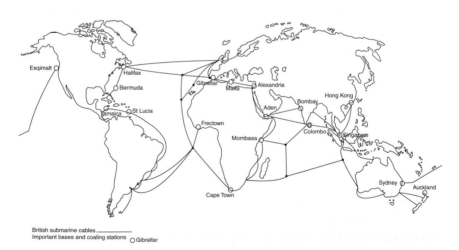

British submarine cables _____
Important bases and coaling stations ○ Gibraltar

FIGURE 2.11 Naval bases and submarine cables of the British Empire, *c.*1900. (Paul Kennedy (2004). *The Rise and Fall of British Naval Mastery.* New York: Penguin, p. 207.)

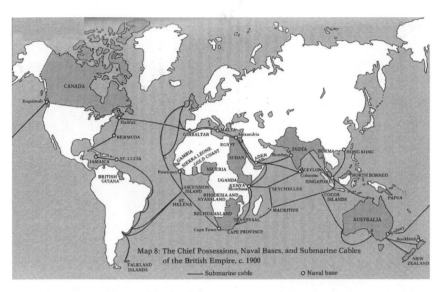

FIGURE 2.12 The chief possessions, naval bases and submarine cables of the British Empire. (Paul Kennedy (1987). *The Rise and Fall of the Great Powers.* New York: Random House, p. 225.)

fuelled by Welsh stoking coal; returning merchant ships brought colonial produce, raw materials such as vegetable oils, grain, rubber and sugar, plus Army regiments on leave. The Royal Navy protected those sea-lanes in both peacetime and wartime (one reason that the Admiralty, despite all its other demands, always insisted upon having such a large number of long-range cruisers).

Only two points about this holistic imperial system can be made in the confines of this chapter. The first is the overwhelming importance of the cable communications network. It was not just that it linked every part of the Empire with each other, but that none of these lines crossed foreign soil or entered foreign harbours. This gave the British an advantage that no other power, even the newer rising empires, possessed, and those trumps would be played to great effect when the First World War arrived. The second was the possession of so many magnificent, fortified naval bases, located at virtually every vital strategic 'choke point' on the globe. The accompanying illustration shows two British capital ships at anchor in the Grand Harbour at Valletta, Malta in 1935, a symbiotic statement of imperial naval power (Figure 2.13). Probably it is the handsome

FIGURE 2.13 *Hood* and *Barham* in Grand Harbour, Valletta, Malta. (Ian Marshall (1990). *Armored Ships*. London: Conway Maritime Press, p. 134.)

warships that catch the amateur observer's eye. Yet even more impressive are the fortifications and docking facilities and communications centres in and around the harbour which made it one of the greatest and (with some Hurricanes and Spitfires for aerial protection) most impregnable naval fortresses in the world. It was not just Malta, though: much the same could be said about Scapa Flow, Rosyth, Dover, Portsmouth, Halifax, Bermuda, Gibraltar, Alexandria, Freetown, Capetown/Simonstown, Aden, Trincomalee, Singapore, Hong Kong and Sydney, as Figure 2.12 shows. If, to return to Polybius, a Roman marching column could feel on familiar ground entering another camp, so, too, could a Royal Naval crew steaming for the first time into Malta or Singapore.

This logistical–communications–intelligence system endured the First World War rather well (what happened on the Western Front and at Jutland is of course another story), and lasted through the 1920s and 1930s, often suggesting – as in that illustration of the British warships in Malta – that little had changed since the 1880s. Even an acute-eyed Briton or foreigner travelling East in, say, 1934, would be struck by the reassurance – the survivability – of it all, as port after port loomed into sight: Southampton first, then Gibraltar, Malta, Cairo, Aden, Bombay, Calcutta, Penang, Singapore, then the 'split' for either China or Australia. Here was the ocular proof.

This system maintained itself even during the Second World War, despite the losses to Japan of much of the British Empire in the East; after all, the Army did hold India, then reconquer Burma in an epic though scarcely known campaign, and finally re-entered Malaya, Singapore and Hong Kong. But thereafter the Empire folded at an incredibly rapid pace, historically compared (i.e. to Rome and Spain). The withdrawal from the Indian subcontinent in 1947–48 – from the keystone to the entire system – was followed in a much swifter time than British officials expected by retreats from Africa, the Caribbean, South-East Asia and the Middle East. Attempts to relocate the 'east of Suez' empire to Kenya or Jordan simply didn't play. The game was over.

Why was this so? The immediate answer must be that that picture of continued British imperial strength represented by the battleships in Malta's Grand Harbour was an illusion. For it was precisely in those years that London was facing the 'overstretch' problem so pithily described by Philip IV's counsellor in 1635. Only the geography was different. The decade of the 1930s brought to the British Empire three serious and parallel challenges: the emergence of Japan as a serious disturber of the status quo across the entire Far East; the disruptive and worrying posturings of Mussolini's Italy in the Mediterranean and Northeast Africa, threatening the shorter British routes to India and the Orient; and, most ominous of all, the resurgence of Germany, this time under a truly vicious and ultimately non-appeasable regime. So, how could Britain dispose of its limited numbers of regiments, air squadrons and warships? If it had a mere fifteen capital ships by the late 1930s (many of them building or being rebuilt), how were they to be disposed – five against Germany, five against Italy, and five against Japan? Or did some other combination of numbers work better? However the ships were deployed, they were never enough. The marvel is that the Royal Navy fought as impressively as it did in the Second World War.

But the larger reason for the Empire's demise was, surely, the relative weakening of the imperial centre's productive system – relative, that is, to other industrialised powers such as Germany, the USSR, Japan and, especially, the United States. It was, after all, from this offshore-island base that ships were built, aircraft assembled, cable lines manufactured and overseas fortresses established. But the heartland had been badly hurt, not just in terms of human loss and psychological retreat but also in its fiscal and commercial endowments, by the Great War. And it was hurt further, and again badly, by the inter-war industrial and trade depression and by the collapse of the international financial system. Those Malta-based battleships, in other words, resembled nothing so much as some earlier portrait of a Castilian platoon shortly before Spain's retreats in the 1640s. And it was not that the British people, and the consolidated peoples of their Commonwealth and Empire, did not fight the good fight between 1939 and 1945. They fought longer and, arguably, harder, than any other combatant. But the sheer material pressures, the changes in the world's productive balances, conspired against them. Thus, it was not that there was diminished vitality in Churchill's Britain; far from it. But it could not last against the vital surge of the world's two superpowers, or the inner fatigue and revolt of the Empire itself. Again, and in retrospect, the real question to be posed is how this four-hundred-year-old, five-continent, seven-sea, single-language artefact maintained itself so long. One doubts if its like will ever be seen again.

The United States

Still, there are many indicators of power which suggest that today's Number One nation, the United States of America, has grounds to claim that it has inherited the position once occupied by Rome, Spain and Britain. In this regard, it is something of a waste of time to debate whether or not the United States is an 'empire'. The real point is: since America possesses a global heft larger than that of any other contender or contenders, how does it sustain itself, and what are the signs that it can maintain – or lose – that position? In seeking to answer those questions, I will follow the previous examples and not devote time to explanations of the rise of this particular great power. It has assumed that the United States possessed masses of relative weight (economic production, strong population, high technology, an efficient military) to reach its present heights, but it is of course unable – it being premature – to declare whether America is at present at its zenith in its trajectory or at the beginning of its relative decline. The evidence is mixed, the debate among intellectuals and the media (especially in the United States itself) is heated, and it seems proper to let readers judge for themselves. My own views on America's longer-term prospects are well known in any case.[1]

Two illustrations in the realms of comparative military power may suffice here. The first is that of the worldwide deployment of US Navy carrier task groups, at least as located shortly before the Al Qaeda attacks upon America of September 11th 2001 (Figure 2.14). This scatter-map may be in fact something of an understatement, in that so many of the carriers were in port and not on the high seas at that time (see remarks in the lower left of the illustration). Nonetheless, this layout affords a glimpse into the global reach of American sea/air power. It suggests, too, immense mobility and flexibility: the USS *Kitty Hawk*'s carrier group was at its home base in Japan at the time of the terrorist attacks upon New York and Washington but, by steaming day and night, it had reached the Persian Gulf during the week following and was deployed forward as a mobile airbase for the aircraft from other carriers which were also racing into the area for attacks upon the Taliban. What all this cannot illustrate, however, is the sheer expensiveness of such power-platforms; for when one adds up not only the building costs for a large nuclear-powered carrier, but also those for its numerous cruiser, destroyer and frigate escorts, plus the costs of the dozens and dozens of aircraft and helicopters, then the total is close to the entire defence budget of Italy. A nation which can afford to pay for thirteen or fourteen such carriers must have deep pockets, or deep debts, indeed.

The second illustration is a more recent map of worldwide US troop deployments. As such, it virtually explains itself (Figure 2.15). What is worth noting is the geographical extent of the newer deployments, putting American Army and Marine units into places that would have amazed Roosevelt or Eisenhower; one is reminded of the remark that the late-Victorian British Army had to grapple with ever more 'crumbling frontiers', moving into a fresh region in order to safeguard an existing but unstable

---

[1] See *The Rise and Fall of the Great Powers* (1987) New York: Random House, chapter 8 and 'The Eagle has landed', *Financial Times*, 1 February 2002.

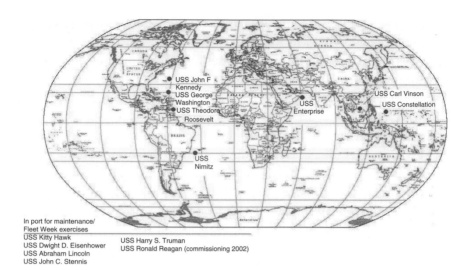

In port for maintenance/
Fleet Week exercises

USS Kitty Hawk
USS Dwight D. Eisenhower
USS Abraham Lincoln
USS John C. Stennis

USS Harry S. Truman
USS Ronald Reagan (commissioning 2002)

FIGURE 2.14   Disposition of US Navy carriers, 1 September 2001.

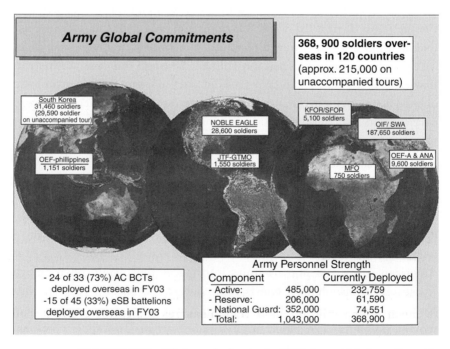

FIGURE 2.15   US Army deployments, 2005. (www.globalsecurity.org)

territory. When these US force deployments occur, they are enormously impressive, at least initially, and certainly no other nation on earth has the airlift capacity to fly a division into central Asia (except, perhaps, Russia) or to put four or five divisions into Iraq. Yet it is also hard to resist the conclusion that these troops are now setting up camp in some of the most insecure and volatile places on the planet; deployments to Germany under NATO command during the Cold War had it far lighter, as things turned out.

What can be witnessed here, then, is an American polity – call it a hegemon, or Empire, or half-Empire – putting its impress upon the world. For better or worse, all roads in international affairs seem these days to lead to Washington DC, as previously they had led to Rome, to the Escorial or to Whitehall. In one particular measure of power – that of military expenditures – the American superiority is even more marked than those of previous empires. In the fiscal year 2005/6, the United States was spending approximately 50% of all global defence expenditures, a proportion that was never achieved by Rome, Spain or Britain in their time of primacy. No wonder American conservatives rejoice, and some of them claim it can last for generations to come, perhaps for the whole twenty-first century. This is their Curzonian (or perhaps Darwinian?) moment.

Obviously, we can have no certainty about the future survivability, or success rate, of the current American superpower, and of whether we are only halfway through the story or somewhat more close to its end – its end, at least, as unchallenged Number One. The previous cases of great Empires described polities that lasted, variously, 450, 150 and 300 years. Will the American pre-eminence last as long as any of them? To borrow from Garvin's original question: 'Will the Empire which dominates the globe today still be as powerful, relative to others, in a hundred years' time?'

Clearly, cautionary language is in order. Our eyes are drawn, naturally enough, to America's present great strengths – the massive armaments, the global reach, the control of world communications. According to so much of the relevant data that measures comparative power politics, America is comfortably ahead, and no challenger is in sight. Yet there are also indicators which suggest that the current superpower may be resting upon far slimmer and more febrile – and thus far less sustainable – foundations than did the Roman and British Empires and that it may not have the staying capacities of even the Spanish Empire.

The arguments which run in the pessimistic direction are usually assembled in the following ways. First of all, there is the inherent unsustainability of the twin deficits – that of the external trade balances, and that of the Federal budget (Figures 2.16 and 2.17). A Number One nation which permits its deficit in the trade in goods and services to plummet so far and so fast is running the risk of future international bankruptcy. And a hegemonic power which occurs such fiscal imbalances, and which at present relies upon half or more of its monthly sales of federal Treasury bonds being purchased by China and Japan is not really, well, a hegemon; and economists' arguments that Asian countries must purchase dollar-denominated securities should make any serious strategist swallow hard. Such ominous tendencies can be, and have been, corrected but not without a strong fiscal discipline that current American political realities and tax structures would find hard to handle.

Secondly, there are the already-existing shifts in the global economic balances that promise to have power-political implications within another generation, if not before (Figure 2.18). If the economists and strategic

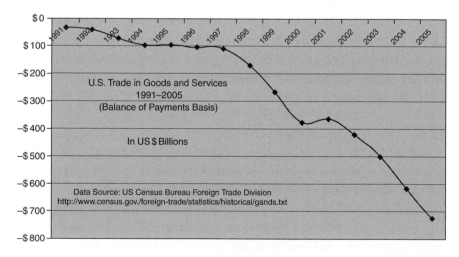

FIGURE 2.16 US trade deficit, 1991–2005. (US Census Bureau, Foreign Trade Division.)

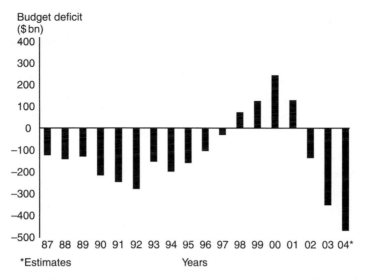

Budget deficit
($ bn)

*Estimates   Years

FIGURE 2.17 US Federal budget deficits, 1987–2004. (Office of Management and Budget; courtesy of BBC News Interactive.)

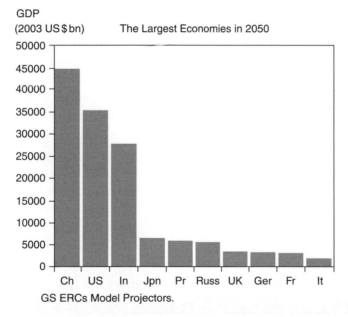

GDP
(2003 US$ bn)   The Largest Economies in 2050

GS ERCs Model Projectors.

FIGURE 2.18 The largest economies in 2050. (Dominic Wilson and Roopa Purushothaman (2003). *Dreaming with BRICs: The Path to 2050*, Global Economics Paper No. 99. New York: Goldman Sachs.)

planners of a serious Wall Street firm like Goldman Sachs can argue that there is some prospect that by the 2050 the People's Republic of China will possesses the world's single largest economy, and that India's total GDP will be close to overtaking America's, what does that do for the Paul Wolfowitz doctrine of the early 1990s of ensuring that the United States stays far ahead of everyone else, forever? The trends (however unlikely to unfold in a straight line) contradict the doctrine.

Thirdly, there exists the prospect that, unless it reduces its overseas commitments, the United States will begin to run out of its Castilian troops, that is, that flow of fresh recruits for the regular US Army, the Reserve and the National Guard, so many of whom come from the American southern states. In other words, the effort of keeping 250,000 US troops in distant, strife-torn parts of the world may be unsustainable in the longer term, even in the medium term, without a change in recruitment policies, or some hard decisions about global priorities. But the first alternative, essentially, a return to some form of conscription, is as politically unpalatable in today's America as it was in Garvin's Britain, where the National Service League's campaign for military service foundered badly. And the challenge of identifying regions of the globe from which the United States can withdraw some of its existing troops, bases and interests is as great as the strategic dilemmas that faced Spain in the 1630s and Britain in the 1930s. One can easily imagine here an updated version of the words of that counsellor of Philip IV in 1635:

> The first and greatest dangers are those that threaten our interests in the Far East, our policies in the Middle East, the preservation of a modern, flexible NATO alliance, and our campaign against world-wide terrorism. Our withdrawal from the Far East would lead to unpredictable consequences in Korea, Taiwan, Japan and elsewhere. A pull-out from the Middle East would shatter our anti-terrorism drive, cause devastatingly high oil prices, and cripple us everywhere else. The collapse of NATO would etc, etc.

Thus, the weary Titan[2] must soldier on, under the too-vast weight of its own fate.

---

[2] The term comes from an address to the 1902 Colonial Conference by Joseph Chamberlain, the Colonial Secretary, where he used the words 'The weary Titan staggers under the too vast orb of its own fate' as he referred to Britain's imperial overstretch following the South African War.

In the grand sweep of history, the coming to the end of America's primacy may not matter all that much (except to arch-patriots in that country), although the manner in which the decline takes place would be of critical importance to everyone in the world. 'If Rome and Carthage fell,' asked Rousseau, 'which Empire then is immortal?' And the answer is none. At present the American Empire, like the Roman, Spanish and British in their heyday, exhibits enormous vitality. But, just like them, the level of that influence upon others is out of proportion to the natural balances of humankind. A country possessing 4.6% of the world's total population cannot indefinitely remain unchallenged at the top, and perhaps especially not in a century of vast technological changes and economic convergence.

Survival of the fittest

This chapter has to attempted to show how Darwin's notion of the 'survival of the fittest' was taken up by certain political circles in Victorian and Edwardian Britain and then employed, to far-reaching effect and with significant consequences, upon imperial and naval policies, and how it also influenced, to a lesser degree, the debates upon economic, social and educational policies. In the second place, and by bringing in a comparative examination of the previous Roman and Spanish empires as well as of the contemporary American superpower, it has also sought to uncover and explain what the author believes to be perhaps the single most important element in the vitality and survival of great powers – namely, the existence of an effective logistical and communications system, superior by far to that of its many rivals. The ghost of Charles Darwin lurks here. Although that renowned scientist refused to be drawn into the political debates about his theory of the survival of the fittest, he surely would have approved of the method of seeking to explain the survivability of great empires by an examination of their strategic nervous systems. In that sense, all of us who study and write upon the grand strategy of nation states must be, to a larger or smaller extent, Darwinists.

The following chapters will explore in turn further strategies we have adopted to ensure the survival of our society and culture. What emerges is a complex system with numerous interdependencies and feedbacks. In the next chapter we continue our explorations by considering the 'survival of culture'.

FURTHER READING

Elliott, J. H. (1963). *Imperial Spain, 1469–1716*. London: Pelican.

Ferrill, A. (1986). *The Fall of the Roman Empire: The Military Explanation*. London: Thames & Hudson.

Garvin, J. L. (1905). 'The maintenance of Empire', pp. 69–143 in Goldman, C. S. (ed.) *The Empire and the Century*. London: John Murray.

Kamen, H. (2002). *Spain's Road to Empire: The Making of a World Power, 1492–1763*. London: Allen Lane.

Kennedy, P. M. (1976). *The Rise and Fall of British Naval Mastery*. London: Allen Lane.

Kennedy, P. M. (1987). *The Rise and Fall of the Great Powers: Economic and Military Conflict from 1500 to 2000*. New York: Random House.

Louis, W. R. (ed.) (1999). *The Oxford History of the British Empire,* 5 vols. Oxford: Oxford University Press.

Mandelbaum, M., (2005). *The Case for Goliath: How America Acts as the World's Government in the Twenty-First Century*. New York: Public Affairs.

Robbins, K. (1983). *The Eclipse of a Great Power: Modern Britain 1870–1975*. Harlow: Longman.

Searle, G. R. (1971). *The Quest for National Efficiency 1899–1914*. Oxford: Blackwell.

Shannon, R. T. (1979). *The Crisis of Imperialism, 1865–1915*. St. Albans: Paladin.

3 Survival of culture

EDITH HALL

Introduction

The 'survival of culture' suggests that as a classical scholar with an interest in the ways that ancient Greek and Roman culture has survived, both until and since the European Renaissance, I should explore examples of this particular type of transhistorical endurance. There are indeed many heart-stopping stories of the close shaves by which some texts and images created by people who lived in the ancient Greek and Roman worlds survived: several, like Longinus's influential treatise *On the Sublime*, have been transmitted in a single, fragile manuscript; others, like Menander's comedies and the Aristotelian *Constitution of Athens*, have been dug up from Egyptian rubbish-dumps on scraps of waste paper, notably the papyri discovered at Oxyrhynchus; uniquely informative images painted on ancient vases were lost in European air raids during the Second World War and can only be studied through the medium of old photographs.

But I am also aware that the title is a hotly contested expression, with two separate connotations that are almost entirely antithetical. On the one hand, the 'survival of culture' has become a slogan of the intellectual right wing. It is, indeed, the title of a recent volume brought out by the two senior editors of the *New Criterion*, a periodical which styles itself, in its website, a 'staunch defender of the values of high culture' and 'articulate scourge of artistic mediocrity'. The editors' collection of essays takes its

Survival, edited by Emily Shuckburgh. Published by Cambridge University Press.
© Darwin College 2008.

cue from Matthew Arnold's definition of culture, in the preface to *Culture and Anarchy*, as 'the best that has been thought and said in the world'. They aim to derogate all the forces that they believe are conspiring to destroy the legacy of culture, by which they mean high *Western* culture – the forces of relativism, postcolonial theory and political correctness.

On the other hand, a goal labelled 'cultural survival' has been espoused by the victims of colonialism and its concomitant homogenisation of culture both past and ongoing. In Australia, the annual Survival Festival Concert is a national celebration of Aboriginal and Torres Strait Islander culture. 'Cultural survival' is a slogan to be found in Native Americans' initiatives aimed at preserving their indigenous languages and customs, and in defences of the Maltese language, a remarkable fusion of Arabic and Italian. Cultural Survival is also the name of the leading US-based international indigenous rights organisation, founded in 1972 and devoted to 'promoting the rights, voices, and visions of indigenous peoples' the world over. In partnership with indigenous peoples, it advocates their human rights before inter-governmental institutions, governments, courts, financial institutions and corporations. It publishes the important quarterly journal also called *Cultural Survival* (Figure 3.1).

So we can't even begin to talk about cultures and survival without parachuting straight in the middle of the Culture Wars. I am going to try to negotiate the semantic minefield by making sure that the work whose survival I explore is actually one of those which most safely arrived in the modern world – the Homeric *Odyssey*. The text of this epic poem was transmitted from pagan antiquity through the Christian Middle Ages in a considerable number of manuscripts, before appearing in print at Florence in 1487. Yet cultural survival requires more than the simple emergence of manuscripts in fifteenth-century Italy: plenty of ancient authors whose manuscripts did so are hardly ever read today. The *Odyssey* as a whole has retained its vigour, rather, because it has been able to establish an unparalleled presence in both popular and elite culture. This chapter explores as a metaphor for our contemporary Culture Wars the famous confrontation between the Greek hero Odysseus and Polyphemus, the one-eyed Cyclops. This episode was already one of the most popular in the visual arts of antiquity (see e.g. Figure 3.2), and has been frequently illustrated and retold since the European Renaissance. Distinguished painters who have

FIGURE 3.1 The cover of a recent issue (Summer 2005) of the quarterly *Cultural Survival*.

been attracted to the theme include Piero di Cosimo as early as 1500, as well as Rubens, Poussin and Fuseli.

In the ninth book of Homer's *Odyssey*, an epic dating to about 750 BC but which preserves far older material, Odysseus (known as Ulysses to the Romans) relates how he arrived at an attractive island, inhabited by the Cyclopes, gigantic beings who lived in caves. He and his crew enter the cave of one of them, Polyphemus, while the giant shepherd is out pasturing his flocks. Odysseus and his men light a fire, and eat some of the Cyclops' cheese. Against the advice of his men, who want just to grab provisions and get away, Odysseus insists on waiting for the Cyclops, in

FIGURE 3.2 Odysseus and some of his men blinding the Cyclops; a painting on a proto-Attic amphora of the seventh century BC, now in Eleusis Museum.

case he can acquire some of the gifts that hosts traditionally bestowed on guests according to an archaic Greek social contract. (Odysseus has, in fact, already failed to respect the terms of this contract by entering the cave completely uninvited.) When the Cyclops returns, he discovers the men in his cave after he has sealed the entrance to it with a boulder, and after some dialogue in which he is told that Odysseus' name is 'Nobody', he begins to eat them, one by one. Odysseus comes up with a plan to inebriate the Cyclops – who is unused to alcohol – and, when he has fallen asleep, to blind him with a stake hardened in the fire (Figure 3.2). The Cyclops tries to call his neighbours to help him, but because he shouts that 'Nobody' has hurt him, they understandably go away. The Greeks escape by hanging onto the underbellies of the Cyclops' farm animals. The episode ends with the Cyclops hurling rocks after the Greeks as they sail away; at this point Odysseus can no longer resist gloating over his feat and announcing his true name, Odysseus, for all to hear. As a result, the Cyclops' father, who happens to be the Sea-God Poseidon, decides to punish Odysseus by giving him a protracted and hazardous journey home. The Cyclops episode thus causes all the remaining adventures in the Odyssey, and is arguably the most famous of them all: David Bader, the irrepressible composer of English-language *haiku*, has recently condensed the *Odyssey* into these seventeen syllables:

> Aegean forecast –
> storms, chance of one-eyed giants,
> delays expected.

Cultural legacy

The *Odyssey* has for centuries had an important role to play in introducing people's imaginations to ancient mythology, through children's versions, of which there have been dozens since Charles Lamb's 1808 *The Adventures of Ulysses* (which was itself much reprinted and which was the text that introduced James Joyce to the *Odyssey*) (Figure 3.3). Until recently all versions have followed Lamb's explicit characterisation of the Cyclops incident, which he upgraded to take centre place in his very first chapter. Lamb's Cyclops is described much more fully and pejoratively than Homer's; he is an 'uncouth monster', with a 'brutal body' and a 'brutish mind'. Lamb also dictated the story's meaning for his juvenile readership by adding that it provided 'manifest proof how far manly wisdom excels brutish force'. Lamb was taking his cue from an ancient intuition that Odysseus's travels somehow symbolised colonial expansion. But before we enter Polyphemus' Homeric cave, to witness perhaps the most famous confrontation between the culturally different in the Western imagination, it may be helpful to survey briefly some of the metaphors that have been invoked to describe our dialogue with what Bernard Knox has ironically called ODWEMS, the 'Oldest Dead White European Males'.

My favourite metaphor for this process has always been the relationship between barnacles and the surfaces to which they attach themselves.

FIGURE 3.3 The cover design to *The Amazing Adventures of Ulysses* (Reproduced by permission of Usborne Publishing, 83–85 Saffron Hill, London, EC1N 8RT, UK. www.usborne.com © 2003, 1982 Usborne Publishing Ltd.)

The metaphor seems particularly appropriate to this book, since Charles Darwin was fascinated by barnacles. He devoted eight years of his life leading up to the publication of on *the Origin of Species* in 1859 to them, writing four whole volumes about them. Barnacles attach themselves to ships, rocks and whales. Barnacle shells obscure the true contours of the ship whose underbelly they cover, thus offering what has always seemed to me a helpful metaphor for layers of meaning which centuries of interpretation, translation and adaptation have affixed to ancient texts. Like barnacles, interpreters are extremely vigorous and often have powerful agendas of their own.

Others have proposed quite different metaphors. The old notion of the 'classical tradition' or the 'classical heritage' takes the idea of a legacy, passed down rather passively down the generations like the family silver teaspoons. For Ulrich von Wilamowitz-Moellendorff, possibly the greatest classical scholar of the late nineteenth and early twentieth centuries, the crucial metaphor of necromancy (the conjuring of the spirits of the dead) came from the scene before Odysseus can talk to the dead in the *Odyssey*: 'the classical scholar has to give his own blood to the spirits of the ancient authors before they will reveal their secrets'. The theatre director Peter Sellars, notorious for his radical topicalisation of the repertoire, sees each classic opera and drama that he directs as an antique house that can be refurnished and redecorated in the style of any era, but still remains essentially the same house. Oliver Taplin proposes the more volatile image of so-called 'Greek fire', a substance used as a weapon by the Byzantines, which discharges a stream of burning fluid that could continue burning even under water. Greek culture is still present in sometimes invisible yet potentially incendiary forms. Rather more ambivalent is Derek Walcott's description of 'All that Greek manure under the green bananas' – the Greek legacy is refuse, indeed excrement, but has also fertilised his Caribbean imagination. This beautifully captures the painful paradox posed by ancient Mediterranean cultural legacy to peoples colonised by the western powers.

Stupid monster

Although the occasional rhetorical controversialist in antiquity decided to speak up for the Cyclops and fashion arguments in his defence, it was almost always from the perspective of Odysseus the wayfarer, the *homo viator*, that the *Odyssey*, including the Cyclops episode, was used in antiquity. This was often in connection with legitimising colonisation and imperial expansion. Odysseus' sons by Circe traditionally founded important cities in central Italy, and the Etruscans were partial to painting scenes from Odysseus' wanderings on the walls of their tombs, as if to assert a cultural ancestry leading back to Greece. At Rome the blinding of the Cyclops came to serve the ideological needs of the Empire. Whoever paid for the spectacular marine picnic resort in the entrance to a cave

FIGURE 3.4 The blinding of the Cyclops: imperial Roman statue group now in Sperlonga Museum.

at Sperlonga (Figure 3.4), where the blinding of the Cyclops was reconstructed on a monumental scale as one of a series of ornamental statues which survived unknown until their excavation in 1957, was buying into the myth of Odysseus the coloniser. The victory over the Cyclops becomes emblematic of the Roman domination of the Ligurian and Tyrrhenian coasts and seas.

By 1400, the Ulysses in Dante's *Inferno* tells how his ships passed through the very Pillars of Hercules, and he exhorted his crew like a prefigurative Columbus, urging them further, across the Atlantic, to what they preferred to think were uninhabited lands:

> O brothers who through a hundred thousand dangers have reached the west . . . choose not to deny experience, following the sun, of the world that has no people.

At the same time, rumours spread of the deformed giants that peopled the new worlds just waiting to be discovered. It was with a potential reality in mind that the accounts of monstrous races on the edges of the world in Pliny's *Natural History* and other ancient compendia were canonised in the medieval period (see Figure 3.5).

It was not just the single eye of the Cyclops that attracted attention: far more important became the Cyclops' role as original man-eater. Reports of cannibalism the world over have been frequently exaggerated. Fears and allegations of man-eating are a standard trope in the xenophobic polemic of nearly every culture and era. The missionaries in Africa may have feared that they might end up in an indigenous casserole, but at least some of the natives believed that the white invader was interested in eating the local people he encountered. Columbus' first letter discusses reports of the Caribbean people of Caniba, terrible cannibals (said in other sources to have only one eye), the image of whom was to play a vital role in the creation of Caliban – the barbarous native just ripe for subjugation – in Shakespeare's *The Tempest*. Columbus also noted how native peoples couldn't handle alcohol, an important element in the presentation of the Cyclops in the *Odyssey* (see Figure 3.6), and subsequently in the literature of colonial encounters between Europeans and their subjects everywhere. The Cyclops then flourished during the great age of teratology, when malformed individuals were studied and feted. This coincided exactly with the first great wave of European colonial expansion, as numerous fabulously illustrated books attest.

By the end of the seventeenth century, when John Locke wrote the second of his *Two Treatises of Government* (1689), the requirements of the expanding mercantilism of the West needed examples of situations in which disrespect of political authority is legitimate. Locke picked on Ulysses' defiance of the barbaric Cyclops' right to govern his own island. Charles Darwin, although appalled by the destruction of some of the indigenous peoples of South America by European imperialists, had certainly himself absorbed the Odyssean myth of the irredeemably barbarous cannibal. In his *Beagle* diaries he expresses his trust in Captain Cook's description of the Cyclopes-like New Zealand natives who throw stones at approaching ships, shouting 'Come on shore and we will kill you and eat you all.' And by the high Victorian Empire of the last three decades of the seventeenth century, the myth of Odysseus' outwitting and subjugation of the Cyclops was deemed an exemplary tale for everyone living under

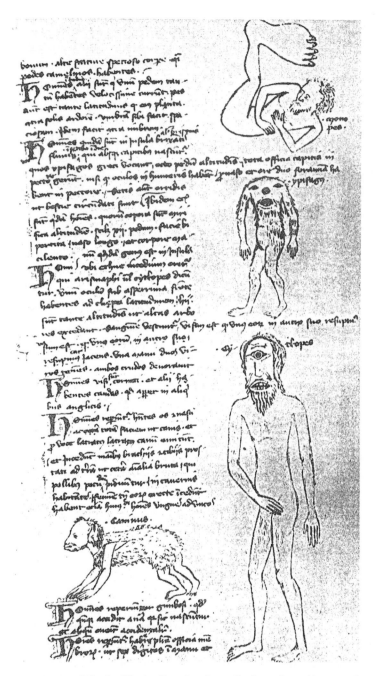

FIGURE 3.5 An illustration showing a Cyclops alongside two other monsters taken from a manuscript illustrating Thomas Cantimprè's bestiary in his thirteenth-century *De Naturis Rerum*. From Friedman, John Block (1981), *The Monstrous Races in Medieval Art and Thought*. (Cambridge, Mass: Harvard University Press). Reproduced with permission.

FIGURE 3.6 John Flaxman, 'Ulysses pouring wine for the Cyclops'. (*The Odyssey of Homer, Engraved from the Compositions of John Flaxman R. A. Sculptor*. (1805). London: Longman.)

Britannia's Rule. In the *Royal Readers*, six standard schoolbooks for educating both children who would be imperial masters, and those who would be their subjects, there was a series of particular heroes and episodes in myth and history selected for their instructive value: they included the Vikings, Napoleon, the Roman Empire, the Conquest of Ireland and, from ancient Greece, Odysseus and the Cyclops.

Big Brother

So far so colonial. The Cyclops thus long represented, in a relatively uncomplicated way, the savages who inhabited shores ripe for invasion and the imposition of civilisation by imperial masters. Yet Immanuel Kant began a new trend to redefine the one-eyed-ness of the Cyclops as narrow-mindedness or tunnel vision. Kant captured this distinction in his contrast between the *sensus privatus* – views that are only formed through privatised or narrow experience, which he called 'Cyclopean thinking', and the *sensus communis*, common or public sense. For Kant, it was entirely possible to be an erudite Cyclops: a Cyclops who knew a good deal about natural history, mathematics, philology or languages. But, without the enlarged thought or public judgement that comes from engagement with a diversity of other viewpoints and perspectives, the learned person fails to think *philosophically*: in Kant's terms, as a member of a living human community.

Closely related to the narrow cognition of the Kantian Cyclops is the widespread association of the one-eyed giant with the Orwellian Big Brother and his sinister, electronic Eye of surveillance. The most familiar example is Stanley Kubrick's intelligent computer HAL in his movie *2001: A Space Odyssey* (1968). But there had long before been self-appointed agents of surveillance of the most real and sinister kind: when the Klu Klux Klan structured itself into the Invisible Empire of the South at a convention in Nashville, Tennessee, in 1867, the group was presided over by a Grand Wizard, with his descending hierarchy of lieutenants, with titles 'Grand Dragons', 'Titans' and 'Grand' or 'Exalted Cyclopses'. The twelfth chapter of James Joyce's *Ulysses* (1922), 'Cyclops', is perhaps the first text critically to identify Cyclopean monocularity with ethnic or nationalist narrow-mindedness. There, sitting in his 'cave' within Barney Kiernan's Dublin pub, is the drinking, belligerent Cyclops; huge and brawny, imagined with a granite spear and 'mighty cudgel rudely fashioned out of paleolithic stone', the obsessive Fenian anti-Semite citizen baits the Jewish hero Leopold Bloom mercilessly.

From the classic novel of Ireland to the classic novel of Black America, Ralph Waldo Ellison's *Invisible Man* (1952). Ellison's Odysseus is transformed into a resourceful and brilliant young African-American man, with a gift for public oratory, seeking his fortune in post-war New York City. The first Cyclopes he meets are in the chapter which centres on his Odysseus-like refusal to give his name to the medical staff subjecting him to electroconvulsive therapy: one doctor inspects him with 'a bright third eye that glowed from the centre of his forehead', and another has 'a circular mirror attached to his forehead'. In the final chapters, however, it is revealed that the autocratic and manipulative white leader of the tyrannical brotherhood – a thin disguise for the Communist Party of the United States in the early 1950s – has only one functioning eye. This Cyclops is particularly sinister because his fanatical and self-serving tunnel vision has masqueraded as a concern for the welfare of the African-American population. In Derek Walcott's stage version of the *Odyssey*, the Cyclops is a totalitarian tyrant who will brook neither dissent nor laughter in his grey police state. And postcolonial theory has associated the Algerian activist Frantz Fanon's work on the privileging of vision by colonial powers with the blinding of the monocular Cyclops: when Odysseus challenges the Cyclops he offers a critique of the inadequate colonial 'look of surveillance', the 'single mode of colonial (super)vision maintained in the evil eye of objectification'. The Cyclops loses the monocular tyrant's power, because he, like the French in Algeria, discovers that monocular, racist vision actually sees 'nobody' at all.

Even Salman Rushdie, somewhat surprisingly, has made use of this Cyclopean tradition in reference to George Bush's government and its demeanour towards the Islamic world. In the aftermath of 9/11, with its masterminds still at large, Rushdie described the United States as:

> a blind giant, flailing uselessly about: like, in fact, the blinded Cyclops Polyphemus of Homeric myth, who was only one-eyed to begin with, who had that eye put out by Ulysses and his fugitive companions, and who was reduced to roaring in impotent rage and hurling boulders in the general direction of Ulysses' taunting voice.

Rushdie contemplates how the episode might reflect Osama bin Laden's own fantasy construction of the global order:

> Polyphemus, after all, is a sort of evil superpower, a stupid creature of great, brute force who respects no laws or gods and devours human flesh, whereas Ulysses is crafty, devious, slippery, uncatchable and dangerous.

The United States is thus framed by Rushdie as running the risk of presenting itself to the rest of the world as the stupid, all-devouring, ugly, blundering giant outwitted by a smaller, clever hero. The flexibility of this type of comparison has meant that the American press has been able to counter this possibility by pointing out that Bin Laden's notorious lieutenant Mullah Mohammed Omar is one-eyed and currently resides in a cave – 'the Cyclops of Al Qaeda'.

The Klu Klux Klan, anti-Semites, white racists, the French in Algeria, US bombardment of Muslim countries – the Cyclops has therefore repeatedly been re-envisioned as the sinister eye of the colonising master, the violent opponent of indigenous or minority cultures. The barbarous primitive on the outer shores of the known world, fit only for subjugation, became transformed, after Kant, into the sinister eye of the racist and the one-eyed totalitarian surveillance machine – thus dialectically turning into the *opposite* of what he had been to the Greeks, Romans, Renaissance explorers and subsequent imposers of global empires. But the story doesn't end here, for the Cyclops has in fact become not only a totemic but a *contested* figure, as Penelope in the *Odyssey* has become for feminists. For there is yet another twist in the tale of the re-envisioning of the Cyclops.

Victim of racism

In 1993, when Davies, Nandy and Sardar published their polemical *Barbaric Others: A Manifesto on Western Racism*, they took it completely for granted that the roles of the good guy and the bad guy in the story of Odysseus needed to be inverted. In their reading, Polyphemus returns to the status of the colonised subject, but now it is as a *victim*, the earliest and most profoundly influential example of 'the analytic categories that swayed the minds of Columbus and his successors'. The Cyclops is here, according to the school of interpretation that sees ancient magic and religion as providing ideological support to ancient Greek civilisation, the colonised and brutally subjected savage of archaic myth. The Stupid Monster who became Big Brother has now turned into a Victim of Racist Oppression.

By 1889, the American historian Henry Adams could imply some sympathy for the Cyclops as colonised subject when he was describing the American reaction to a British naval attack on the unprepared USS *Chesapeake* in 1813:

> The brand seethed and hissed like the glowing olive-stake of Ulysses in the Cyclops' eye, until the whole American people, like Cyclops, roared

> with pain and stood frantic on the shore, hurling abuse at their enemy, who taunted them from his safe ships.

But the most important factor in the rehabilitation of the Cyclops was the twentieth-century, post-Freudian reappraisal of Odysseus. He first became a problematic, repressive and violent figure when the Polish dramatist Stanislaw Wyspianski wrote his tragic play *Return of Odysseus (Powrót Odysa)* in 1907, and Gerhart Hauptmann countered with *The Bow of Odysseus* in 1914. These central European Odysseuses have been traumatised by their war experience and have become psychotically wedded to violence for its own sake, unable to take up the roles of responsible father, husband or leader.

Such critiques of Odysseus inevitably led to the re-evaluation of his victims. In philosophy the major assault on Odyssean heroism came when Theodor Adorno and Max Horkheimer (central to the Frankfurt School of neo-Marxist social theory) collaborated on *Dialektik der Aufklärung*, first published in New York in the dark days of 1944. In an increasingly irrational and intractable world they sought to activate every conceivable resource made available through philosophical reasoning. Simultaneously they identified the deadly role played by *reason* in the creation of humankind's problems, at least in the form of means – end calculations and the specious objectivity of ideologically motivated science. When they set out to trace the genealogy of the dark underbelly of Western reason, it was the voyage of Odysseus which they selected for their allegorical case study. Thus they traced the destructive potential of reason not to the usual suspect – the much-maligned eighteenth-century Enlightenment – but to the *Odyssey*, one of the very earliest surviving charter texts of Western culture. They argue that this Odyssean rationality, already bound to identity, inevitably represses, cancels and tramples on singularity, difference and otherness. Reason offers humans extraordinary, unhoped-for success in dominating nature through scientific and intellectual advancements. But it inevitably leads to the domination of some men by others, and of most women by most men.

For the Frankfurt School, Polyphemus the Cyclops, in his ideal pastoral existence, is the creation of an imagination already racked with sorrow at its alienation from the environment and at its own self-imposed policing of the disposal of time. Polyphemus, they recognise, becomes the model 'for the evolving line of stupid devils of the Christian era right up to Shylock and Mephistopheles'. Most importantly, they argue that the very stupidity of the giant comes to represent 'something better as soon as it is subverted by the one who ought to know better'. They are the first explicitly to recognise that Odysseus *abuses* his intellectual powers on the Cyclops' island – that he is trespassing with all the arrogance of a colonial master, and creates a situation which can only result in barbarism and bloodshed. The dialectic of Enlightenment means that Odysseus cannot assert his superiority without dialectically beginning to behave as badly as his supposed inferior. This is always the conundrum of Empire – that its justification (the ascent from primitive barbarism) is inevitably cancelled by the physical or cultural violence required to impose it. Even more importantly, the critical theorists see that the minute Odysseus behaves this badly, the *stupidity* of the Cyclops begins to look more like benign naivety.

Alterity

Closely allied with the Frankfurt School's reading is a the last reason for the recent vigour of the Cyclops as cultural presence: the proposition that his eye is no more nor less than a marker of radical *difference*. For this group of interpreters, the Cyclops represents the way that ancient Greek colonisers *imagined* the different types of human that they encountered on their marauding voyages. Their own different appearance, diet and mode of production, and the fear they feel become projected onto the figure of the primitive ogre and crystallised in him. What is now needed, it is being argued, is not identification with the Greek adventurer as he invades the home of the Cyclops, devours his food, intoxicates him, tricks him and blinds him – a triumphalist celebration of the superiority of cunning over physical size and of the Greek's right to subdue and dominate – but a reading that tries to imagine what it *felt like to be the Cyclops*, that turns him into the *subject* of the narrative, rather than its *object*.

This line of argument owes much to the readings of Polyphemus' close relation Caliban in Shakespeare's fantasy *The Tempest*, the colonial agenda of which was pointed out by Aimé Césaire in his 1969 version of the play. The position has been taken to its furthest limit by Sylvia Wynter, who proposes that black readers should practise what she calls 'a Cyclopean poetics of reading'. She argues that the Cyclops defines radical difference within the repertoire of images encoded in Western culture, on the level of 'marvellous reality'. She deliberately offers a counter-mythology to lineages previously proposed by other black, postcolonial and anti-racist writers, claiming that it is impossible to square the contradiction between Odysseus' relationship with white colonisation and the black poets' desire to identify their own quest for freedom with Odysseus' pursuit of his goals. For Wynter, the problem is that if you side with Odysseus, you *inevitably* end up conspiring in the binary oppositions that have figured as 'Other', of 'a population group or human hereditary variation (the Negro)'. Wynter therefore makes explicit what has been implicit in some black writing previously, that the time has come to embark on 'the as yet still unexplored nature of *what must be* the quest of the Cyclops', a new poetics which has as the goal of 'its Cyclopean quest... the assumption and revalorization of the being and perspective of *alterity*'.

There have been attempts to address this contradiction. Historians of popular culture have suggested that the mutant X-Men, invented by Marvel Comics in 1963 (Figure 3.7), were a covert or unconscious fantasy exploration of the civil rights crisis. The X-Men's physical difference (their alterity) leads them to face state-sponsored racism, bigotry, prejudice and lynchings. Their leader, Scott Summers, is known as 'Cyclops' on account of the visor he must wear to protect his powerful eyes. Cyclops' parents are enslaved; the mutants' patron, Professor X, has been likened by different commentators both to Martin Luther King and to Malcolm X. In *Sea Grapes* (1976), Derek Walcott grappled with the contradiction by suggesting that the Cyclops was actually involved in *creating* the epic *Odyssey*, giving primacy to the blindness the Cyclops shares with Homer, the author of the epic: 'the blind giant's boulder heaved the trough / from whose groundswell the great hexameters come / to the conclusion of exhausted surf'. It is with the image of the Cyclops' boulder that Walcott reminds us that *nothing* – not even the glories of ancient Greek poetry – can ever compensate adequately for the pain of cultural loss and dispossession. In Walcott's *Omeros*, too, Odysseus is fleetingly linked with the clever European persecutor of the black children of the Caribbean, themselves

FIGURE 3.7 The first issue of *The X-Men* (1 September 1963). © 2006 Marvel. Used with permission.

associated with the Cyclops' flock, in stories 'we recited as children lifted with the rock / of Polyphemus'.

Instances of the Cyclops becoming a point of identification by oppressed ethnic groups, or anti-colonial polemicists, can now be multiplied; he has recently been used to make the case for the Hawaiian movement for independence. But the new Cyclops can be traced much further back, to a passage in Aimé Césaire's prose poem *Return to My Native Land* (1939).

This Martiniquan intellectual trained as a teacher of Latin and Greek at the prestigious École Normale Supérieure, and the poem relates his quest for identity in Paris and back home. In one episode the author–narrator, a black man who has become an Odyssean victor-figure in Paris, encounters on a tram another, larger black man, whose eye socket has been hollowed by Poverty. Wynter names this stunning section *Encounter with the Cyclops on a Paris Tram*:

> And I, and I,
> I who sung with clenched fist
> You must be told the length to which I carried cowardice.
> In a tram one night, facing me, a Negro.
> He was a Negro tall as a pongo who tried to make himself very
> small on a tram seat. On that filthy tram seat he tried to abandon
> his gigantic legs and his starved boxer's trembling hands. And
> everything had left him, was leaving him. His nose was like a
> peninsula off its moorings; even his negritude was losing its
> colour through the effects of a perpetual tanner's bleach. And the
> tanner was Poverty...
> He was an ungainly Negro without rhythm or measure.
> A Negro whose eyes rolled with bloodshot weariness.
> A Negro without shame, and his big smelly toes sniggered in the
>   deep gaping lair of his shoes.
> Poverty, it has to be said, had taken great pains to finish him off.
> She had hollowed the eye socket and painted it with a cosmetic
>   of dust and rheum...
> And the whole thing added up perfectly to a hideous Negro, a
> peevish Negro, a melancholy Negro, a slumped Negro, hands
> folded as in prayer upon a knotty stick.
> A Negro shrouded in an old, threadbare jacket. A Negro who
> was comical and ugly, and behind me women giggled as they
> looked at him.
> He was COMICAL AND UGLY
> COMICAL AND UGLY, for a fact.
> I sported a smile of complicity...

This is devastating stuff. Indeed, the final speech Césaire gives to Caliban in his anti-colonial adaptation of *The Tempest* could equally well have issued straight from Polyphemus' mouth, if instead of 'Prospero' we insert the name 'Odysseus':

71

Odysseus, you are the master of illusion.
Lying is your trademark.
And you have lied so much to me
(lied about the world, lied about me)
that you have ended by imposing on me
an image of myself.
Underdeveloped, you brand me, inferior,
that's the way you have forced me to see myself.
I detest that image! What's more it's a lie
But now I know you, you old cancer,
and I know myself as well.

The survival of classical culture

Stupid monster, Big Brother, or Victim of Racism – it is by contemplating the experience of our Cyclops over the last hundred years that some possible ways may be suggested through the apparent impasse created by the two different understandings of cultural survival. If what passed between Odysseus and Polyphemus in that cave can have so many urgent and wholly conflicting reverberations in our time, perhaps the story's very susceptibility to plural interpretations offers a potential answer to the seemingly insoluble claims of the survival of culture, defined traditionally, and cultural survival, defined anthropologically. In the first place, it is now clear that contemplating the widest possible range of interpretations of the ancient Greek and Latin classics is the surest method of ensuring their continuing vigour and survival in the third millennium. At least three excellent books by classical scholars including François Hartog (1996), Irad Malkin (1998) and Carol Dougherty (2001) have recently approached the *Odyssey* as a protocolonial fantasy. A corollary of this is the belief that a fresher, and more perceptive, account of the contents of the original texts can be rendered if it is informed by the most recent responses to it – and especially the most radical and surprising among them.

If after travelling with the Cyclops through time we now go back to Homer's text, for example, it will be with heightened sensitivity. We recognise that although Odysseus alleges that the Cyclopes have no laws, and no social bonds between households, Polyphemus' neighbours do in fact run to help Polyphemus when he calls on them, and indeed operate a system of mutual support and hospitality. We will also learn to be wary of translators, many of whom, convinced that they understand the moral outlines of the story, have often imposed meanings of their own: one influential version introduces the term 'race' where Homer simply says 'the Cyclopes'. The phrase that introduces Polyphemus is usually translated something like 'a *giant* who used to pasture his flocks far afield, alone' (9.187–8), but the Greek actually says 'a huge *man*' (*anēr pelōrios*): the word for 'man' here (*anēr*) elsewhere moreover defines men *in opposition* to monsters (see 21.303); the term *pelōrios*, 'huge', is also used in Homer of such glamorously big heroes as Ares, Achilles, Ajax and Hector.

The Greek marauders enter uninvited, light a fire and eat food. The Cyclops does not notice their presence until he has returned home and sealed his cave in order to keep his flocks inside. When he realises what is going on he does eat Odysseus' men, and indeed constitutes a threat to the lives of all of them, but (cannibalism apart) is this form of self-defence really so shocking? Odysseus, after all, later kills the suitors who exactly mirror his own actions in the land of the Cyclopes, since the suitors are also intruders who sit uninvited at another man's hearth and consume his livelihood. Moreover, the Texas Penal Code article 9.42, 'Deadly Force to Protect Property', argues that it is justifiable to kill a trespasser on your property when

> the actor reasonably believes that deadly force is immediately necessary to prevent the other's imminent commission of arson, burglary, robbery, aggravated robbery, theft during the night-time or criminal mischief during the night-time; or to prevent the other who is fleeing immediately after committing burglary, robbery, aggravated robbery.

Classical scholars have now had to reread the *Odyssey* and accept that, at least in a Texan court, the Homeric Cyclops would today have a watertight defence!

The Cyclops, as part of classical culture, may originally have become popular because he provided, on the level of fantasy, a justification of colonial expansion and of empires like those examined in Paul Kennedy's chapter of this book. But in the twentieth century, he began to flourish because of something quite antithetical – the liberation of colonised, oppressed, exploited and threatened peoples. Polyphemus and his struggle with Odysseus have had such wholly conflicting reverberations in recent times that they can be read as a symbolic paradigm of the struggle over the classical canon. Their showdown metaphorically represents both the conflicting views about the contents of the canon suitable for a postcolonial age, and the ways in which those contents should be read. There are still defenders of elite culture who insist that the *Odyssey*, as an ancient Greek masterpiece, is somehow as inherently superior to most non-European literature as Odysseus was to his monstrous victims; whereas, some black critics see no possibility that Odysseus, as protocoloniser, robber and assailant of Polyphemus, can be recuperated as a hero by anyone sensitive to the history of racism. Is there any way through this impasse? How can enjoying the 'Western Classics' be compatible with opposition to Western imperialism and cultural or racial oppression?

The first way of responding to this question comes from an unconventional classicist Norman Austin who pondered on the difference in tone between the Cyclops episode and the rest of the epic, noticing, correctly, how childlike the two leading characters are (which partly explains their enduring fascination for children). Odysseus wants lots of presents like a child; Polyphemus is a playground bully; they bicker and squabble and brag. Unlike most of Odysseus' adventures – Calypso, Circe, Nausicaa – this one offers no grown-up erotic interest, nor even palace coup: two men–boys slug it out, to the point of death and mutilation, over a few dairy products. Indeed, Austin proposes in a reading that owes much to Melanie Klein that what the Cyclops' cavernous dairy represents is the womb and the breast, and that what we are facing here is the most regressive and infantile sibling rivalry. Not only does Austin's psychoanalytical discussion apprehend the *tone* of the story better than anything I have read, but in the postcolonial global village, the notion of squabbles between brothers under the skin can perhaps help the survival of *all* cultures, indigenous, Western, pre-Christian pagan and non-Western alike. The myth represented in the *Odyssey* belongs to everyone and no one. Narrow-mindedness, childishness and sibling rivalry know no ethnic boundaries.

Of the numerous painters attracted to the Cyclops episode in the *Odyssey*, many have portrayed the actual incident of the blinding. But the defiant anger of Polyphemus as he hurled great rocks at the escaping ship has also featured, above all in Joseph Turner's awe-inspiring Romantic oil painting *Ulysses Deriding Polyphemus: Homer's Odyssey* (1829), in the National Gallery; here Polyphemus looms over Odysseus' galleon like a configuration of dark, angry clouds, an elemental force of nature. It was with typical perceptiveness that John Ruskin commented that 'Polyphemus asserts his perfect power', so that this must be 'considered as the central picture in Turner's career'. Jean-Léon Gérôme subsequently took a rather less mystical approach to the same episode, but his painting is interesting because the viewer is invited to see the casting of the boulder very nearly from Polyphemus' point of view (Figure 3.8). This seems to be the sort of perspective taken in an ancient Armenian story about a giant, and looking at other myth systems provides a second possible answer to the problem presented to the global village by the values that have previously been attached to the classical canon. The Armenian hero Turkʿ Angeleay, who has forebears as early as Mesopotamian myth, bears some striking similarities to Polyphemus, including great size and some special quality to his eyesight. Yet for the Armenians Turkʿ was always a celebrated figure precisely because he fought off wicked pirates, brigands and looters, who sailed too close to his country's coastline, by tossing huge

FIGURE 3.8   *Polyphemus* (*c*.1902), by Jean-Léon Gérôme.

boulders at them. An eminent scholar of Armenian myth, James Russell, has recently written of Turk⁽:

> When we stand with him on the Pontic shore as he hurls boulders at a pirate ship, it may come as a shock that some Odysseus is aboard: this is the Anatolian mirror-image of Greek heroism, from an Anatolian source . . .

The *Odyssey* represents just a single cultural expression – however influential it has been – of a far more ancient set of stories shared by cultures wherever *Homo sapiens* has ever travelled. The type of the Cyclops figure is manifested in a very wide range of myths and folk tales that have been recorded the world over. One etymological explanation of Polyphemus' name is that it means 'speaking many languages' or 'spoken of in many languages': large mythical shepherds have always transcended cultures quite as much as clever mythical travellers. The monocularity of the pastoral giant, living cheek-by-jowl with his flocks and herds, was probably a response not to anything remotely to do with ethnic difference, but with a universally occurring genetic abnormality that all primitive humans will have observed occasionally amongst the premature fetuses aborted by their sheep and cows.

Yet a third and surely the most important response to the problem faced by the modern world, as it attempts to disentangle the Western classics from the terrible legacy of Empire, must come from the new strategies being developed in the work of contemporary 'transcultural' writers. The Cyclops story certainly appeals to Wilson Harris, born in Guyana (then called British Guiana) in 1921. His parents combined Amerindian, African and European blood, and he objects to being ethnically categorised. He also refuses to be forced into a choice between rejecting or embracing literary traditions of any kind simply because of the contingent values that they have historically embodied. In *The Mask of the Beggar* (2003), Harris fuses the *Odyssey* with the pre-Columbian Aztec figure of Quetzalcoatl in order to ask whether humans can find spiritual ways to transcend their tragic history of mutual barbarism through stressing the threads that connect, rather than divide, their imaginative lives. For Harris, the labile figure of the Cyclops sometimes represents the innocence of the pcoples massacred by the conquistadors, but at others the blindness of societies that are still today imprisoned by what should have become long outdated hostilities. Similarly, the Maori poet Robert Sullivan (Figure 3.9) draws his ancestry both from the very Nga Puhi people of New Zealand's North Island of whom Darwin was so terrified, but also from Galway in western Ireland. He has thought long and hard about how to negotiate the Scylla of conservative advocacy of the survival of culture and the Charybdis of threats to the cultural survival of indigenous

FIGURE 3.9   The poet Robert Sullivan.

peoples. Sullivan has recently used the myth of Orpheus to explore the legacy of Western imperialism in the south Pacific in his libretto *Captain Cook in the Underworld* (2002), where the opening chorus absolves Cook – directly associated with Ulysses – for the temerity of his claim to have 'discovered' a land that could not be discovered because it was inhabited already:

> Forgive the Ulysses
> of his day, for the mores of his age,
> for overlooking the inhabitants with his claim.

But Sullivan has also imbued with intense Odyssean resonances a collection of poems about Maori seafaring, *Star Waka* (1999) (a 'waka' is a canoe). These offer the reader a jumble of voices that explore the contradictions within the indigenous New Zealanders' relationship with the Western canon. One voice is able to acknowledge the bravery of the poor European settlers who sailed to New Zealand 'over the edge of the world / into Hades / the infernal Greek and Latin-ness of many headed creatures'. Another angrily derides Odysseus for depriving him of his rightful place in the poem and subjecting the Maoris to the curious stare of anthropologists instead of respecting the brilliance of Maori stellar cosmogony. And yet, to be heard within this complicated polyphony, is another calmer, more reflective voice offering a message about culture and its now inevitable globalisation that seems both resigned and somehow more hopeful and forward-looking:

> Do not mind the settler. I observe
> The rules of this mythology (see how he did not
> place a star or ocean or a waka
> in his pageantry). I am Odysseus,
> summoned to these pages by extraordinary
> claims of the narrator. I run through all narratives.

ACKNOWLEDGEMENTS
Thanks for help in the research and completion of this article to Mary Beard, Peter Gathercole, Geoffrey Lloyd, Philip Lofthouse, Margaret Malamud, Polly Weddle, Richard Poynder and especially Emily Shuckburgh for her exceptional care and diligence as an editor.

FURTHER READING

Austin, N. (1983). 'Odysseus and the Cyclops: Who is Who?' in Rubino, C. A. and Shelmerdine, C. W. (eds.) *Approaches to Homer*. Austin: University of Texas Press, pp. 3–37.

Farrell, J. (2004). 'Roman Homer', in R. Fowler (ed.) *The Cambridge Companion to Homer*. Cambridge: Cambridge University Press, pp. 254–71.

Hall, E. (1989). *Inventing the Barbarian: Greek Self-Definition through Tragedy*. Oxford: Oxford University Press.

Hall, E. (2008). *The Return of Ulysses: A Cultural History of Homer's Odyssey*. London: I.B. Tauris & Co. Ltd.

Hardwick, L. (1996). *A Daidalos in the late-modern age? Transplanting Homer into Derek Walcott's* The Odyssey: *A stage version*, online article available at http://www2.open.ac.uk/ClassicalStudies/GreekPlays/conf96/cctoc.htm.

Highet, G. (1949). *The Classical Tradition: Greek and Roman Influences on Western Literature*. London: Oxford University Press.

Knox, B. M. W. (1993). *The Oldest Dead White European Males: And Other Reflections on the Classics*. New York: W. W. Norton.

Lloyd-Jones, H. (1982). *Blood for the Ghosts: Classical Influences in the Nineteenth and Twentieth Centuries*. London: Duckworth.

Reid, J. D. (1993). *The Oxford Guide to Classical Mythology in the Arts*. Oxford: Oxford University Press.

Soyinka, W. (2002). 'Beware the Cyclops', *Guardian Saturday Review*, 6 April, 1–2.

Wallace, M. O. (2002). ' "What ails you Polyphemus?" Toward a new ontology of vision in Frantz Fanon's *Black Skin White Masks*', in his *Constructing the Black Masculine: Identity and Ideality in African American Men's Literature and Culture, 1775–1995*. Durham: Duke University Press, pp. 170–8.

Wynter, S. (2002). ' "A different kind of creature": Caribbean literature, the Cyclops factor and the second poetics of the *propter nos*', in T. J. Reiss (ed.) *Sisyphus and Eldorado: Magical and Other Realisms in Caribbean Literature*, 2nd edn. Trenton, NJ and Asmara, Eritrea: African World Press, pp. 143–67.

4 Survival of language

PETER AUSTIN

Introduction

When I was in New Mexico in 1980 I heard Navajo fluently spoken by
Native American people of all ages. Now, twenty-five years later, many
of the Navajo population still speak Navajo at home, but fewer and fewer
children are learning it as their first language. The Navajo language, and
with it essential elements of Navajo culture, are becoming endangered.

Gamilaraay (or Kamilaroi) is an Australian Aboriginal language, from
the north-west of New South Wales. By the 1940s local Aboriginal social
and cultural transmission had been so disturbed by the impact of European
settlement that the language had ceased to be used as the main means
of communication. Knowledge of words and expressions (such as plant,
animal and food names) continues to this day, but by the 1960s no fluent
native speakers of the language remained. Today however the language is
being revived and reintroduced in local schools. Beginning in the 1990s,
there has been intense local interest in the language, and strong support
for its documentation and reintroduction.

There are about 6700 languages spoken on Earth today, each closely
tied to the cultural identity of the speaker communities. A threat to the
survival of the language of one of these communities is, in the terminology
of the previous chapter, a threat to their 'cultural survival'. To consider

Survival, edited by Emily Shuckburgh. Published by Cambridge University Press.
© Darwin College 2008.

this threat I will take as my starting point a provocative quotation on the back cover of a recent book by Andrew Dalby:

a language dies every two weeks: what are we going to do about it?

Are languages indeed disappearing and, if so, how and why? Where do suggestions come from of language loss at the rate of *one language per fortnight*? What are some possible responses to what looks like an impending crisis for the survival of the world's languages? Are all smaller languages doomed to extinction, or are there signs that loss of languages can be reversed?

Language diversity

The world's languages can be ranked in terms of the estimated size of the populations who habitually use them as a first (or native) language, and as a language of wider communication (typically spoken as a second or third language). It is difficult to determine speaker numbers exactly, and the very notion of 'speaker' is somewhat vague, but Table 4.1 gives estimates (see also Raymond Gordon's *Ethnologue: Languages of the World*) which show the relative positions of the world's top ten languages, all of which have over 130 million speakers.

Some languages, such as Spanish, Russian, French and Arabic, can be considered multinational since they are used and officially recognised in a number of countries; English is a truly global language now present across the world and increasing both in use and prestige. The top ten languages are spoken by 40% of the world's population, and the top twenty are spoken by half of the people on Earth.

In fact, 96% of the world's population speaks just 4% of the 6700 languages, which means that only 4% of the world's population maintains 96% of its linguistic diversity.

There are thus a few very large languages with many millions of speakers, and very many small languages with a few thousand, or a few hundred speakers (and in the case of indigenous languages in Australia and North America dozens of languages spoken by just a handful of people each, or in some instances, by just one person). The geographical

Table 4.1 *Top ten languages in terms of speaker numbers*

	Language	First-language speakers	All speakers	Distribution
1	Mandarin Chinese	1.10 billion	1.12 billion	
2	English	330 million	480 million	Global
3	Spanish	300 million	320 million	Multinational
4	Russian	260 million	285 million	Multinational
5	French	75 million	265 million	Multinational
6	Hindi/Urdu	250 million	250 million	
7	Arabic	200 million	221 million	Multinational
8	Portuguese	160 million	188 million	
9	Bengali	185 million	185 million	
10	Japanese	125 million	133 million	

Table 4.2 *Distribution of indigenous languages by area*

Area	Languages		No. of speakers
	Count	Percent	
Africa	2000	30	675 887 000
Asia	2200	33	3 489 897 000
Americas	1000	15	47 559 000
Europe	200	3	1 504 393 000
Pacific	1300	19	6 124 000
Totals	6700	100	5 723 861 000

Source: Based on Gordon (2005) (see www.ethnologue.com).

distribution of languages is also quite skewed, with the richest diversity in Africa, South and South-East Asia and Latin America, as shown in Table 4.2.

Probably the most linguistically diverse places in the world are to be found in the Pacific: the island of Papua New Guinea with a population of 3 million has 1100 languages, while Vanuatu's 100 000 people speak an incredible 120 languages (this being the greatest density of languages per head of any country). Moreover, these diverse languages are frequently radically different from one another, belonging to entirely different language *families*. Although most European languages belong to the Indo-European family, Papua New Guinea has at least 40 distinct families. Great diversity in terms of language families can also be found in the American continent, where South America counts some 93 distinct language families and North America some 50 distinct families.

Extinction of language

Economic, political, social and cultural power is in the hands of the speakers of the large languages. The many thousands of small languages are marginalised and under pressure from the larger ones. In the past sixty years, from the end of the Second World War onwards, there has been radical reduction in speaker numbers of smaller indigenous languages, especially in Australia and the Americas. In addition, as with the Navajo, speaker communities show increasing age profiles where older people continue to speak the languages but younger ones do not and have shifted towards the few larger multinational languages. Sometimes this takes place rapidly, over a generation or two, often via a period of unstable multilingualism. Sometimes the language shift is gradual, but inexorable, and occurs over several generations.

Beginning in 1990, linguists such as Michael Krauss of the Alaska Native Language Center rang alarm bells, suggesting that in the twenty-first century up to 90% of human languages would become extinct. Krauss's predictions are perhaps extreme; however most scholars now agree that at least half of the world's linguistic diversity will disappear over the next 100 years: this means a loss of more than 3000 languages forever. At this rate we arrive at Dalby's estimate of one language lost per fortnight, a figure far in excess of predictions of loss of species diversity for endangered plants and animals.

It is possible to identify a number of factors involved in this loss of language diversity and to develop a typology of speaker communities. An important factor is *intergenerational language transmission*, that is, whether or not children are learning the language from their parents and care-givers. A not uncommon situation is for parents to speak a heritage language among themselves in private and to converse with their children in a large language of wider communication that is socially, politically and economically dominant (this may be a multinational language such as Spanish or French, but it can also be a dominant regional language such as Wolof, Hausa or Swahili, to mention some African examples). In the 1950s and 1960s, many Welsh parents would speak Welsh in private, but would speak English to their children and the intergenerational transmission weakened. Low transmission inevitably leads to language shift towards the dominant tongue.

A second factor is *percentage of speakers* among the total population, that is, not the total number of people speaking a language but the proportion of a given community who continue to use it. Language loss is associated with reducing proportions of the population using the language. Note that small languages with even 500 or 1000 speakers (as in the Asia–Pacific region) are not currently undergoing shift because virtually all the community uses them on a daily basis while other larger languages such as Quechua (which has millions of speakers in South America) are being lost as increasing numbers of younger people are dominant or monolingual in Spanish.

A third parameter is *domains and functions of use*, that is, the contexts and situations where the language is regularly used. Some languages are restricted just to the family domain for personal communication between friends and relatives (with a dominant language being used outside the house), while other languages show a wider range of contexts and uses, including education, religion, trade and business, and government. Constriction of domains and functions can lead to language loss, particularly when dominant languages begin to encroach into the domains previously reserved for use of smaller languages as a result of young people switching to the spreading language and bringing it with them into the social and family sphere.

A fourth and very important parameter is the *attitudes and language ideology* of the community, and of their neighbours. An ideology that values multilingualism and variety is less likely to lead to language loss than one that sees monolingualism as normal and multilingualism as problematic or threatening (to local or national social and political cohesion). Frequently the economically, politically, socially and culturally powerful speakers of large languages are firm believers in the benefits of mono- rather than multilingualism.

Speaker evaluation of their language is another factor: communities who positively value their language as an expression of their culture and identity are typically less likely to give it up than those who negatively evaluate their way of speaking and stigmatise it as an unwritten (or even unwriteable), a 'dialect' rather than a full language, ugly, or not worthy of learning by outsiders.

Examination of these factors enables us to develop a typology of language situations. Some languages can be described as *viable*, *safe* or *strong* languages. These languages are spoken by all age groups, with a very wide distribution in the community, with high intergenerational transmission, actively supported and positively evaluated. Chinese, English, Spanish, Russian and French are in no danger, nor are any of the other top 4% of the world's languages mentioned above.

A second group is *endangered* languages, typically spoken by socially and economically disadvantaged populations, under pressure from a larger language, used by a reducing proportion of the population, and usually not being intergenerationally transmitted, that is adults are not passing them on to children in large numbers. Endangered languages are under threat of loss unless their current contexts of use and acquisition change.

A third category is *moribund* languages, namely those no longer being learnt by children at all, used by reducing numbers of older speakers with very little social function in highly restricted domains. Maybe there are just two old men who get together once a week to tell jokes to one another and that is the only use the language will have. Moribund languages die as the remaining speakers age and pass on.

Finally there are *extinct* languages with no native speakers and no usage. Among the indigenous languages of Australia, for example, over 60% are now extinct (with another 35% moribund and just 5% viable). One of the languages I worked on in Australia, an Aboriginal language called Jiwarli, became extinct in May 1986. It was traditionally spoken along the upper reaches of the Henry River, a tributary of the Ashburton River, in the north-west of Western Australia. I know the exact date it became extinct because I worked with the very last person who spoke that language, Jack Butler.

Within speaker communities there are several types of speakers that can be identified. *Fluent speakers* control a wide range of styles and have extensive vocabulary and lexicon – they can say anything they want to appropriately in any context. For endangered and moribund languages we also often find *semi-speakers* who have partial control over the language and show gaps in their knowledge of the grammar, lexicon and usage. They may be able to understand fluent speakers perfectly but are themselves limited in their active competence. Finally, we can encounter *rememberers* who do not use the language actively but can recall words or expressions used by an older generation of fluent or semi-speakers who they heard using the language when they were children. Rememberers often are unable to form grammatical sentences in the language or to recall more than a handful of everyday words. As languages change their status over time from safe to endangered to moribund to extinct we typically find reductions in fluent speaker numbers, increasing semi-speakers and finally only rememberers.

Causes of language loss

We may ask: why are languages becoming endangered? Why do speakers not use them any more? There are two major reasons: external and internal causes. External causes are outside the control of the community themselves and are typically associated with military, religious, political, cultural, economic or social and educational subjugation. This may arise from colonialist policies (understanding colonies in the broadest sense) of outside powers, or from internal colonialist government policies aimed at suppressing linguistic diversity. There are also external causes such as physical and medical catastrophes which result in population loss: disease, natural disasters and famine, which are discussed in following chapters; all may contribute to languages becoming endangered. The tsunami that hit the north coast of Papua New Guinea on 17 July 1998 obliterated villages around Sissano lagoon. Similarly, the devastating tsunami of 26 December 2004 almost wiped out communities living on the Andaman Islands who speak several Andamanese languages belonging to a distinct language family, unrelated to languages found outside the islands. HIV/AIDS, which is decimating populations in Africa, is another catastrophic example. These latter causes are much rarer than subjugation however.

Internal causes are those triggered by the community's negative evaluation of their language and culture (often arising itself from external subjugation) and positive valuation of some other language which appears to be a key to access to wider opportunities; together these result in lack of transmission to the younger generation. The combination of external and internal causes is leading to language loss globally. Historically speaking, this is not a new development; languages have been lost throughout history, and probably during prehistoric times as well. For example, in 100 BC there were dozens of languages spoken around the Mediterranean that rapidly disappeared with the spread of Latin and Greek promoted by the Roman Empire, such as Etruscan, Oscan, Elymaic, Venetic and Dacian (see Andrew Dalby's book). For most of these only fragmentary records remain. The difference now is that in the late twentieth and early twenty-first centuries new phenomena have emerged, including the development of nation–states with monolingual ideologies. The communications and interconnections, which as we saw in Paul Kennedy's chapter are so critical to Empires, are now global. The result is that the process of language loss is taking place all over the world at a pace never seen before.

When Europeans first arrived in Australia in the eighteenth century, there were hundreds of groups speaking different languages, each of them with their own customs, their own way of speaking. There were probably a million people in some 600 different groups. These figures are rough because for many areas we just don't have the data. Those people spoke 250 different languages, 250 Australian Aboriginal languages, with two major groupings in the north and the south, and possibly twenty-five different families. Within each community, each group had different ways of speaking. A man would have had to speak to his mother-in-law in a completely different way to the way he spoke to his father or his brother or sister. There were special styles used for ritual and song. And there was widespread multilingualism through exogamous marriage. Typically men married women from another group, exogamously marrying outside the group, so children grew up in situations where their mother and father spoke totally different languages. An aunt and an uncle would speak a further one or two different languages and their grandparents something else as well. Frequently Aboriginal people spoke four, five, six, ten different languages, *completely* different languages.

Today, for the vast majority of Australia, the languages are gone. Only twelve languages of the original 250 languages remain viable. There is bilingual education in some areas, but it is under attack from local governments. The Northern Territories government recently overturned bilingual education and removed funds from twenty programmes teaching languages. There is strong Aboriginal cultural identification, but languages are rapidly being lost amongst younger age groups. The children are switching to English or to a creole version of English, a language that is impossible for English speakers to understand but is based on the contact between English and Aboriginal languages. And for many Aboriginal people there are pressing social problems like health, drug and alcohol abuse, housing, and horrendous unemployment patterns. Those place huge burdens that go well beyond a concern for linguistic heritage.

Saving languages

Should this rapid loss of linguistic diversity be a cause for concern? For many people, especially those living in countries which promote monolingualism, the answer is 'no'. Language diversity is seen as a problem creating divisions and conflict: the fewer languages the better. Unfortunately, this view is both naive and has many counter-examples: all sides in the troubles in Northern Ireland spoke English, as did the combatants in the American Civil War. Speaking the same language does not guarantee communication and peace. In addition, all languages are subject to change over time so that two populations originally speaking the same language but geographically separated and with different histories can even over a few generations see their ways of speaking diverge to the point of unintelligibility. Difficulties of communication between British, American and Australian English speakers, for example, are symptomatic of the inexorable divergence that time brings (not to mention Indian English, Singapore English or Jamaican English where influences from other local languages and colonial histories are apparent).

An alternative view is that we *should* care because linguistic diversity is a good thing, just as environmentalists argue for ecological diversity. It is not clear however that there are direct parallels between plant and animal species on the one hand and languages on the other: language is culturally not genetically acquired and humans can be multilingual (whereas species identity is unique). A stronger argument is that for many communities languages express *identity*. Language is not just about communication: we need languages to talk to and trade with our neighbours, but equally importantly we use language to express our identity, to express who we are, to differentiate ourselves from our neighbours, and to group ourselves with the people that we consider to be part of us. The emergence of three separate languages in former Yugoslavia (Bosnian, Serbian and Croatian) in place of Serbo-Croatian is a reflection of this issue of identity along with a number of other social and political factors.

In addition, languages are important as repositories of history and culture. Every language and every speaker community in the world has oral history and oral culture: songs and stories, mythology, records of the ancestors, stories of where people came from, how they came to be, how they live their lives, what their world is. Language encodes and carries this knowledge. If it is lost, if the language disappears, so too is culture, history and knowledge lost. (This argument is promoted strongly by UNESCO and other international bodies under the banner of 'intangible cultural heritage'.) Indeed sometimes knowledge from oral tradition can prove critical to the survival of the community itself: there are numerous examples of communities apparently responding appropriately to unusual natural disasters as a result of advice passed down through the generations in oral tradition, as we shall see in a later chapter. We can also argue more broadly that linguistic diversity is important because language is part of collective human knowledge. Each language represents a different way of viewing the world, each is a different way of thinking about and talking about the world. This is evident in the problems with translation between different languages, and difficulties with matching words, idioms and expressions between languages. Linguists additionally would argue that linguistic diversity is important scientifically since languages are the stuff upon which our understanding of pronunciation, grammar and vocabulary is based, and ultimately, what it is that makes humans unique as the only species that uses language. Since most languages have not been documented, let alone well described, it is essential that the problem of language loss be addressed from a linguistic perspective.

So is it a hopeless situation with 50% or more of the world's languages disappearing because of language shift? There is growing evidence that the answer is 'no', because language shift can be reversed. For example, the 2001 Welsh language census shows a rise in the number of speakers compared to ten years previously, especially in southern Wales. In addition the number of children speaking Welsh today is higher than it has been for a generation because of the spread of bilingual education and other support measures; the raising of Welsh political and cultural consciousness has meant that the language has undergone a resurgence. A second example is Maori of New Zealand which has seen an increase in speaker numbers via *kohanga reo* 'language nest' pre-schools which bring fluent Maori speaking grandparents into contact with child learners in a Maori-only environment. Children of the language nests established twenty years ago are now undertaking Maori medium education at tertiary level. A similar model has been adopted for reintroduction of Hawaiian, which had almost disappeared as a spoken language but is now seeing increased speaker numbers, especially among those learning it as a second language. In Guatemala there is now a Mayan Language Academy, and similar examples of revival and revitalisation can be given around the world (see Mark Abley's book for other instances); we return to this topic below.

If we decide that for all these reasons the survival of languages is important and we want to keep languages being spoken and used, what can we do about it? How can we bring about the survival of language? People outside of communities where languages are under threat can work in a number of areas: language documentation, language protection, and language support, including *revitalisation* (giving vitality back to languages by extending the domains and functions where they are used). This work must be carried out in a collaborative and respectful manner with members of the speaker communities if they choose to do so. Collaboration with, and respect for, the community is essential because while some communities are keen to try to maintain their cultures and languages, others actively want to give up their language and past and to assimilate to the dominant society and culture. To promote the survival of languages we need to understand the patterns of use and the attitudes of the speaker communities. These considerations are crucially important because they determine whether a language will manage to continue, and they provide reliable and comprehensible information for developing intervention strategies.

Language documentation involves collecting linguistic, socio-linguistic and cultural data, including audio, video and text materials to create a corpus which can then serve as a resource to be used by educators and others. It is also important to collect information about the social, cultural and political environment of the community; we need to understand the 'ecosystem' from a linguistic perspective in order to understand the processes of language shift. Documentation needs to be properly archived with a trusted language repository, along with relevant metadata (such as who is speaking, when, where, how, under what circumstances, for what purpose) so that materials are widely accessible both currently and into the future. Professional language archives are now being established in a number of locations around the world.[1]

Outsiders can also become involved in community education, helping communities to understand the circumstances of their language, its vitality or lack of it. Very frequently we find in threatened language situations that a particular family group or community does not understand the broader perspective of the trajectory of their language. For this reason it is important that linguists like myself make our research materials available to the speaker communities and not just to an academic linguistic audience. We also need to provide research training and teacher training opportunities to members of the communities so that skills can be transferred and community members will have the required expertise to act on their wish to see the language continue to be used. Just as with the problems of disease, famine, natural disasters and climate change to be discussed in the following chapters, there also needs to be involvement in policy issues to help support the survival of languages. Monitoring regional and national language policies is important to ensure, for example, that a country such as Nigeria with more than 500 languages supports this

[1] See www.delaman.org for a list.

diversity. Institutional support for training and policy development needs to be provided, as does support for students to work on languages, technical and other facilities, and to help with the production of books and other materials

Economic and political support is important because there is a direct relationship between economic, political and health survival and the retention of language and culture. Poverty, poor health, discrimination and lack of political power are among the conditions that give rise to language and cultural shift and language loss. Attacking the root causes of the shift supports communities who want to retain their languages and cultures but who feel they cannot because of pressures on young people such as migration to urban centres for work, education, health and political reasons. This is particularly the case in Africa and Asia where we find the massive linguistic diversity mentioned above. Publicity is also a significant challenge. By disseminating information about all aspects of the problem as widely as possible both within the speaker communities themselves and outside we can raise awareness of endangered languages inside and outside communities where they are spoken through all the available channels (see the list of 'Further reading' at the end of this chapter for some examples).

The last five years have seen a number of significant developments on all these fronts, including major initiatives in language documentation, policy-making, international publicity and language support, as well as emergence of grass-roots programmes. In 2000 the Volkswagen Stiftung initiated its DoBeS (*Dokumentation der bedrohte Sprachen*, 'Documentation of endangered languages') project and currently supports forty teams of researchers across the world documenting, archiving and supporting endangered languages. Figure 4.1 shows the research sites of the DoBeS project teams. In 2002 the Lisbet Rausing Charitable Fund through a benefaction of £20 million established an Endangered Languages Documentation Project at the School of Oriental and African Studies (SOAS) in London that is currently funding seventy research project teams across the world.[2] Figure 4.2 shows the main locations where work is progressing. In addition, there is an Academic Programme at SOAS which trains postgraduate students in language documentation and support, and an

---

[2] See www.hrelp.org.

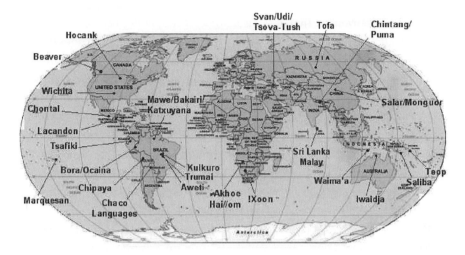

FIGURE 4.1 Research sites of DoBeS-funded documentation of endangered languages project teams.

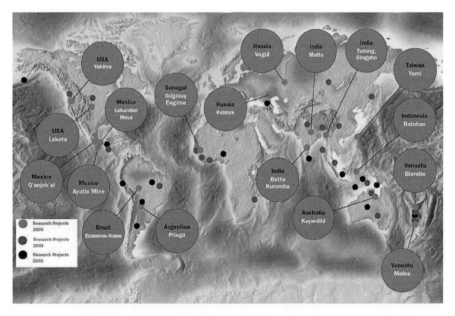

FIGURE 4.2 Main locations of SOAS Endangered Languages Documentation Project research teams.

archiving programme that is establishing a major digital archive repository of all the materials collected by the researchers and students. The SOAS project also publishes books, multimedia CD-ROMs, and trains and supports a wide range of researchers, students and community members. Other initiatives include the Documentation of Endangered Languages project of the National Science Foundation and the National Endowment for the Humanities in the United States, the Endangered Languages Programme of the Netherlands Scientific Organisation, the Endangered Languages Fund (based in the United States), the Foundation for Endangered Languages (based in the United Kingdom) and UNESCO through its intangible cultural heritage and education work.

Scores of grass-roots community-based initiatives have also appeared in recent years, including many which draw upon the Maori and Hawaiian models, as well as master – apprentice schemes that team a fluent native speaker together with a younger language learner.[3] Information technology is increasingly playing a role in these initiatives, with growing use of multimedia. In Australia, Aboriginal language centres that are run and operated by indigenous people are a major force in language documentation and support, especially in remote areas. Tribal Colleges in the United States and Canada take an active role in language education. Information sharing through regular workshops is also a feature of this kind of language documentation and revitalisation. In California, 'Breath of Life' workshops bring together older speakers of Californian languages with younger people to 'breathe life' back into languages that are endangered.

At the start of this chapter I mentioned the Australian Aboriginal language Gamilaraay. I have been involved with this language for over thirty years and as a case study it can illustrate how collaboration between outside specialists and community members can play a role in language survival. The Gamilaraay and Yuwaalaraay languages were traditionally spoken in the north-west of New South Wales, over a vast territory (see Figure 4.3) from the Great Dividing Range near Tamworth, north and west to the Darling and Barwon Rivers. There was a range of dialect variation within this region, mostly marked by vocabulary differences, with all local groups identified as *gamil* 'no' *-araay* 'having'. The socio-linguistic history of Gamilaraay is typical of many south-eastern Australian languages.

---

[3] See, for example, http://babel.uoregon.edu/nili.

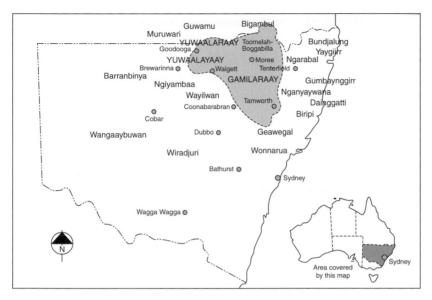

FIGURE 4.3 Location of Gamilaraay and Yuwaalaraay (based on Ash, Giacon and Lissarrague, 2003).

In the 1850s when Europeans began to settle in this region Gamilaraay and Yuwaalaraay were the main languages of the Aboriginal population; however, by 1900 the languages were in retreat. The Gamilaraay and Yuwaalaraay were driven off their land and exterminated, shot and poisoned. Their waterholes and camping grounds were fouled, and they died from infectious diseases brought by the settlers against which they had no immunity such as influenza, smallpox and venereal disease. Reserves were established and the remnant populations forced together to live in squalid conditions where only English was allowed as the language in public. Children were taken from their parents and put into dormitories in schools away from the reservations under a policy of assimilation, resulting in what is now known as 'the stolen generations'. Shockingly, these policies only ended in the 1970s.

By the 1950s, one hundred years after first contact, Gamilaraay and Yuwaalaraay were private languages used only by old people in houses and in camps, while young people were all speaking English. When I first started working in this area in 1972, I encountered only semi-speakers and rememberers, and no fluent speakers remained. Most people with some knowledge of the language could only recall expressions their parents and grandparents had used. By combining their knowledge with recordings made in 1955 by Professor Stephen Wurm of the last fluent speakers, and early settler and missionary writings, it has been possible to document the core vocabulary and grammar of these languages (see Ash, Giacon and Lissarrague, 2003).

In 1988, the bicentennial of the settlement of Australia prompted a major national discussion about the settlement process and the loss of Aboriginal languages and cultures. In 1992 the High Court Mabo judgment overturned the fiction that Australia was *terra nullius* ('unoccupied land') when claimed by the British Crown, and led to the restoration of some land rights to indigenous people. On 10 December 1992 Prime Minister Paul Keating admitted:

> We took the traditional lands and smashed the traditional way of life. We brought the disasters. The alcohol. We committed the murders. We took the children from their mothers. We practised discrimination and exclusion.

Interest in language and culture also accompanied these changed sentiments; a small dictionary of Gamilaraay that I published in 1992 was reprinted four times, such was the demand for copies. In 1993 I worked with the a local Gamilaraay man, the late Bill Reid, to develop materials and publicise language issues, and in 1995 David Nathan and I developed an online dictionary and thesaurus of Gamilaraay which was the first fully hypertext bilingual dictionary on the internet. This was launched after extensive community consultation and support and has been widely used. By 1994 local linguist John Giacon had helped to establish a language programme in primary school and adult education evening classes, and community-initiated meetings began to be held. In 2002–03 a stream of language support materials was published including books and CDs with titles like *Gamilaraay–Yuwaalaraay Guwaaldanha Ngiyani* ('We are speaking Gamilaraay–Yuwaalaraay'), *Yaama Maliyaa* ('Hello Friend') and *Yugal* (Gamilaraay and Yuwaalaraay Songs). The last of these includes pop music sung in the indigenous languages. Merchandise such as T-shirts with Gamilaraay words on them, and signage in schools was developed, extending the domains of the language and bringing it out into the public arena. After extensive lobbying, the New South Wales government approved teaching of Gamilaraay and Yuwaalaraay (and other Aboriginal languages from the state) as 'Language Other than English' curriculum in primary and secondary schools, and in 2004 provided a significant injection of funds to develop more materials and curriculum. This recognition has assigned significant status to both the languages and the educational programmes.

Today Gamilaraay and Yuwaalaraay are in use in a school context, and have developed important iconic functions such as for music, greetings and speech-making at public events. Whether usage extends further remains to be seen, but it is clear that after a long period of neglect and almost certain death, the languages have undergone revival as a result of outside support, grass-roots activity and official recognition.

Future challenges

So the loss of a language can be reversed. Measures can be put in place that bring about the survival of a language – or at least that prevent its immediate demise. But there remain a number of urgent challenges for work on endangered languages. Recruitment, training and capacity-building for language documentation and support are major obstacles to be overcome. Documentation and language maintenance principles have been emerging over the past ten years but we need to understand better the theory and practice of these new areas. And there is an urgent need to communicate with the linguistic community and the wider world about the state of languages and their futures. Within the global village that is so dominated by a few large languages, how many people are aware of the extent of language diversity? How many people are aware that about 6700 languages are spoken on Earth today, but that many of these are endangered? In my view, progress can be made only through community involvement, and achieving the right balance between knowledge and skills in multidisciplinary teams involving outsiders, insiders, community activists, government workers, anthropologists, linguists, scientists and information-technology specialists. Finally, there is the task of raising consciousness among the general public and in endangered language communities, combating the notion that monolingualism is a 'natural state of affairs' and that the world would be better off if only everyone else spoke the same. Multilingualism should be seen as a boon, not a problem. Speaking a language of wider communication can be achieved by *adding* to one's linguistic repertoire, not by subtraction.

It is only through progress on these measures that language endangerment can be confronted. Language is intimately linked to our identity; not just to our identity as individuals, but also to our identity as a species. There are at least 3000 languages that are in need of support. It is imperative that we act before it is too late so we can assure the survival of the languages and all that is associated with them.

FURTHER READING

Abley, M. (2003). *Spoken Here: Travels among Threatened Languages*. London: Heinemann.

Austin, P. and Nathan, D. (1998). *Kamilaroi/Gamilaraay Internet Dictionary*. http://coombs.anu.edu.au/WWWVLPages/AborigPages/LANG/ GAMDICT/GAMDICT.HTM

Ash, A., Giacon, J. and Lissarrague, A. (2003). *Gamilaraay, Yuwaalaraay and Yuwaalayaay Dictionary*. Alice Springs: IAD Press.

Crystal, D. (2000). *Language Death*. Cambridge: Cambridge University Press.

Dalby, A. (2002). *Language in Danger: How Language Loss Threatens our Future*. London: Penguin.

Gordon, R. G., Jr. (ed.) (2005). *Ethnologue: Languages of the World*, 15th edn. Dallas: SIL International. (www.ethnologue.com)

Nettle, D. and Romaine, S. (2000). *Vanishing Voices: The Extinction of the World's Languages*. Oxford: Oxford University Press.

5 Surviving disease

RICHARD FEACHEM AND OLIVER SABOT[†]

Introduction

The previous chapter described the threat posed to us, humankind, by the loss of language and argued the case for action to promote the survival of language. In this chapter we will discuss the threat of disease and in particular the two greatest current and potential global pandemics: HIV/AIDS and adapted avian influenza. In following chapters, further threats to humankind from famine, natural disasters and climate change are examined, and there are of course many more potential threats that will not be discussed here. But before we continue with our journey, we will first consider the thorny issue of at which threat should we the human race direct most attention?

My big scary global threat is more important than yours

The intellectual game called 'my big scary global threat is more important than yours' has been popular for decades, possibly for centuries. In this game, those concerned with global well-being seek to persuade each other, and a broader audience of intellectuals and opinion formers, that a particular hazard to global order is of special significance and importance

[†] Executive Communications Officer at the Global Fund to Fight AIDS, Tuberculosis and Malaria.

Survival, edited by Emily Shuckburgh. Published by Cambridge University Press.
© Darwin College 2008.

and requires extra effort, resources, and attention. Over the past decade, an annual rhythm for the playing of this game has evolved around the summits of the G8 nations (Canada, France, Germany, Italy, Japan, Russia, the United Kingdom and the United States) and their preparatory phases. Lobby groups advocate for their particular causes. Pundits hold forth in the opinion pages of the serious newspapers. And the national interests of G8 members are advanced through numerous meetings and lengthy negotiations among the so-called G8 Sherpas.

None of the causes being advanced is unworthy or unimportant. They include climate change, a plethora of other environmental concerns, energy security, nuclear security and non-proliferation, world trade, terrorism, the Millennium Development Goals and a variety of other global challenges and good causes. In an ideal world, all of these would receive adequate attention and appropriate international mechanisms to move the various agendas forward would be agreed and implemented. The reality is different. Financial resources and political space are finite. If a G8 Summit pays more attention to the need for a new HIV vaccine (as in 2004), some other topic will receive less attention. If a G8 Summit focuses especially on development in Africa (as in 2005), other important topics will be to some degree marginalised. Trying to get the priorities right and to determine whether indeed 'my big scary global threat is more important than yours' therefore becomes important.

It has become popular in recent years to use the concept of a global public good as a means to establish high priority. It is commonly argued that a global public good has a special call on the attention of the global or international community because global public goods can only be provided through concerted international action. Let us pause briefly, therefore, to consider what a global public good is and is not, and how this may relate to the main topic of this chapter and indeed to this book as a whole.

Clarifying global public goods

The term 'public good' is much bandied around today in the media and lay discourse. It has come to mean roughly 'something that is good for the public'. For economists, of course, public good is a *terminus technicus* and has a precise meaning. A 'public good' is one that exhibits the features of non-rivalness and non-excludability. A car is not a public good. A car has rivalness. If I buy a car, you cannot buy the same car and the supply of all cars available to you for purchase is reduced by one. A car exhibits excludability. I can prevent you from riding in my car and I can also prevent you from free-riding in my car. If I choose to let you ride in my car, I can ensure that you pay me for the privilege. By contrast, law and order, which is the archetypal public good, exhibits in full non-rivalness and non-excludability. If I benefit from law and order, this does not reduce the quantum of this good available for the benefit of my neighbours or my fellow citizens. Similarly, the governments who pay for law and order (using general taxation raised from all citizens) cannot exclude any citizen from the benefits of law and order and cannot levy charges on individual citizens in relation to their degree of benefit from the lawful and orderly environment in which they live.

This is an important concept because a public good will always exhibit a high degree of market failure and will always, therefore, require a high degree of government intervention to ensure that it is financed and provided to a socially optimal extent. Note the language *ensure finance and provision* rather than *finance and provide*. To directly finance and provide will always be an option, but it is not the only option for the effective supply of public goods.

A 'global public good' is a public good for which the properties of non-rivalness and non-excludability apply between and among countries. Thus one country's consumption does not reduce the supply available for other countries. And if one or a group of countries choose to purchase a certain global public good, all other countries can free-ride and cannot readily be excluded from the consequent benefits. Nuclear non-proliferation is a good example. By contrast, girls' education, while undoubtedly being a global priority and a global good thing, is not a global public good. Female literacy in Malawi, for example, is achievable irrespective of female literacy in Zambia and, if achieved, provides few benefits for Zambia.

Just as it requires government intervention to ensure the finance and provision of public goods to an optimal extent, global intervention is required to ensure the finance and provision of global public goods. It is useful to imagine a government of the world. Suppose we had a government of the world with a minimalistic 'Jeffersonian' view of its roles and responsibilities. Such a government of the world would focus solely on ensuring the finance and provision of global public goods and would go to great lengths to prioritise these, and to concentrate only on the few highest priorities. Jefferson himself best described this mandate:

> a wise and frugal government, which shall restrain men from injuring one another, which shall leave them otherwise free to regulate their own pursuits of industry and improvement.

At the top of the list of priorities would surely come global law and order, just as law and order comes at the top of every national list of the duties of government. There would be high competition for second place, but our choice would be preparing for and controlling global pandemics. The reasons for this choice are twofold. First is the strictly global public good nature of global pandemic control. The corollary of this is that, without concerted global action, there will not be preparedness and there will not be control. There is no alternative. Second is the magnitude of the consequences that follow from a failure to be prepared and a failure to control. In other words, the size of the global public bad. Look only at the case of HIV/AIDS. Already, 25 million people have died. Today, 40 million people are infected. The economic and social consequences are massive and are still growing. This catastrophe has completely run away from us. We were not prepared and we have failed to mount an effective response in the first twenty-five years of the known pandemic. As discussed below, reasonable estimates of the scale of the devastation which may be caused by a new influenza virus are far larger and, without question, the pandemic will overtake us far more rapidly (Figure 5.1).

The nature of global pandemics

The spread of infectious diseases among countries and among continents is not new. It has been well documented from ancient times and has been commonplace over the past six centuries. Examples include the spread through Europe of bubonic plague (the 'Black Death') and cholera and the

FIGURE 5.1 It is not whether but when the next pandemic will occur.

introduction by Europeans to the Americas and the Pacific Islands of such infections as smallpox and measles.

The world today is significantly different from even a hundred years ago and the biological possibilities for global pandemics are now considerably more restricted than they used to be. In the days when all human beings all over the world lived, for the most part, in situations of poverty and poor hygiene, any infectious disease arriving newly in a country could take hold and spread. Infectious diseases are spread, in the main, by one of five principal mechanisms:

- by the faecal–oral route which allows pathogens contained in the intestine of one person to be ingested by another person;
- by the respiratory route in which a pathogen is exhaled by an infected person and inhaled by another person;
- by sexual transmission, through heterosexual or homosexual contact between two persons;
- by insects, as in the case of bubonic plague or malaria; and
- by medical or pseudo-medical interventions, as with tattooing, circumcision, surgery with less than perfect hygiene, vaccination with reused needles or syringes, various transfusion practices, and so forth.

The two most documented examples of pandemics sweeping through Europe in the past few hundred years, bubonic plague and cholera, had transmission linked to insects (plague) and the faecal–oral route (cholera). Both were strongly linked to urban squalor and poverty. Neither of these is easily possible today either in rich countries or in more wealthy communities in poor countries. The conditions of poor hygiene, water supply and sanitation which allow the rapid spread of cholera are not conditions now found commonly in wealthy countries or wealthy communities. Although there are isolated cases of cholera in the southern United States every summer, cholera outbreaks do not occur. Similarly, for insect-borne diseases, although isolated malaria cases occur in Europe and the United States, outbreaks do not occur because the conditions that would sustain them no longer exist. Truly global pandemics have therefore become restricted to infectious diseases whose mode of transmission are, primarily, respiratory and sexual.

At the same time, the range of disease agents which are capable of causing global pandemics has also narrowed. Medical science has armed us with compounds which can eliminate or mitigate most infections by three of the four primary agents: *bacteria* (cholera), *protozoa* (malaria) and *helminths* (schistosomiasis). Diseases caused by these agents are now largely confined to the developing world where populations do not have ready access to the necessary medications and other interventions. Viruses, on the other hand, continue to confound the best efforts of the scientific establishment and are thus the main causative threat for global pandemics. It is therefore no surprise that the devastating global pandemic which is upon us, HIV/AIDS, is a sexually transmitted virus and that the devastating global pandemic which we fear, influenza, is also viral and has a respiratory mode of transmission.

Yet even as the possible avenues and agents for global pandemics have been restricted over the past century, the potential devastation of those pandemics that do occur has increased dramatically. There are three primary reasons for this. Firstly, in order to cause a pandemic in the modern environment, a pathogen must be capable of bypassing the host of defences humans have constructed against them, from vaccines and other medical interventions to hygiene and quarantines. Those few that do make it past these obstacles are highly resilient and adaptable, able to exploit our remaining weaknesses to rapidly and efficiently proliferate among humans – a classic application of Darwin's famous theorem. The second cause for the greater potential impact of modern pandemics is that there are simply more people to kill. At the time of the last great global pandemic, the so-called 'Spanish Flu' of 1918–19, global population was roughly 1.8 billion. Today, it is 6.5 billion. And, lastly, that increased global population is intermixing with much greater speed and frequency than ever before. At the turn of the last century, it took roughly ten weeks to travel from Hong Kong to London, and even this was a marked improvement from the days before the steam engine and the Suez Canal. In 1950, the same trip took three days on a lumbering airliner that made stops in nine different European and Asian cities along the way. Today, it takes thirteen hours. As a result, a new virus that emerges in East Asia can spread to every corner of the globe within a day. At the same time, this greater ease of travel has also caused the volume of human traffic to increase significantly. For example, in 1950, roughly 1 million people travelled in and out of the United Kingdom by air. This year, the National Air Traffic Services expects to handle flights carrying more than 180 million people. Pathogens therefore have ample opportunity to travel rapidly among their human hosts.

The case for pandemics

To return to our earlier assertion, we must now prove that pandemics are indeed deserving of a high priority slot for action by the global community. Surely, sceptics will argue, in this age of miracle medicine, infectious diseases cannot rival the potential devastation of a nuclear explosion or a significant rise in sea levels. We therefore examine in detail the human and social impact of the two greatest current and potential global pandemics: HIV/AIDS and adapted avian influenza. Before doing so, it is worth noting that while these diseases rightly attract the most attention, they are not the only pandemic threats currently faced by the world. Tuberculosis and malaria together kill 3 million people every year, as many as HIV/AIDS, though they are largely confined to the developing world. And the emergence of severe acute respiratory syndrome (SARS) in 2003 served as a reminder that the causative agent of the next great pandemic may not yet exist.

The current pandemic: HIV/AIDS

HIV/AIDS emerged during a heady time in infectious disease control. Smallpox, perhaps the most feared disease over centuries of human history, had been successfully eradicated. Three other ancient scourges – malaria, polio and tuberculosis – had been banished to the developing world and, in the case of polio, eradication was in sight. And antibiotic 'magic bullets' had transformed syphilis and other sexually transmitted diseases into mere nuisances.

It is partly due to this budding hubris that HIV/AIDS was at first met with such indifference by leaders throughout the world. For years, the response of the US government was paltry funding and confident statements that a vaccine was imminent. Another less flattering explanation for the desultory response throughout the world was the marginalisation of those initially most affected: homosexuals, intravenous drug users and immigrants. Leaders in the United States and other developed nations eventually mustered the necessary leadership and resources in time to slow and partially contain the epidemics in their countries. But it has not been until the last five years that a concerted response has been mounted at the global level.

It is too little, too late. In the two and a half decades since the first known HIV/AIDS case, 65 million people have been infected with the disease. More than 25 million of them have died. And the disease continues to spread rapidly across the globe, with more than 5 million people newly infected every year. Sub-Saharan Africa has been the heart of the pandemic to date, accounting for nearly two-thirds of global infections and hosting countries where as many as two out of every five adults are living with the disease (see Figure 5.2). But it may not hold that dubious honour for long. India may have already surpassed South Africa as the country with the greatest number of infections and, if not, will do so soon. China and Russia are experiencing rapid growth of the disease. Unmitigated, generalised epidemics in these three giants, home to two-fifths of the world's population, would far exceed anything witnessed to date.

In Africa, the pandemic is taking a toll not just on individuals and communities, but on the welfare of entire nations. Unlike many other infectious diseases, HIV/AIDS primarily strikes down adults in their most

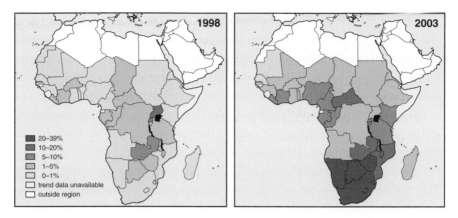

FIGURE 5.2   Prevalence of HIV in adults in sub-Saharan Africa, 1988–2003.

productive years. The impact of a full-blown epidemic therefore extends into every sector of society. Families lose their breadwinners and care-givers, companies their employees and managers, and governments their soldiers, civil servants, and leaders. In Zambia, for example, the number of teachers dying each year account for roughly two-thirds of new recruits entering the education system. Absenteeism and death of employees is significantly increasing labour costs for South African companies, reducing their competitiveness in the global economy. The effect on the broader business environment in high prevalence countries has been described as 'running Adam Smith in reverse'. Instead of growth leading companies to increasingly specialise their workforce as Smith described, the pandemic is forcing them to scale back training and other investments in human capital as more and more of their current and potential employees succumb to the disease.

The economic impact of the HIV/AIDS pandemic is massive, though the broad and often intangible effects of the disease, as well as the dearth of data available in most developing countries, make it difficult to fully capture the aggregate impact of the disease on national and regional economies. Initial studies have suggested that the pandemic is respon-sible for a 2-4 % reduction in the annual GDP of African countries. These are conservative estimates. A study by the World Bank and Heidelberg University is more ambitious in its attempts to capture the long-term impacts of the disease such as the compound loss of human capital over generations as children are increasingly denied the benefits of education and parental guidance. It concludes that, if current trends continue, South Africa, which accounts for a quarter of Africa's aggregate GDP, will experience a complete economic collapse within three generations.

Even this more comprehensive model cannot fully measure the impact of the pandemic on the basic social fabric in countries hit hard by the disease. An estimated 15 million children have already been orphaned by the disease and millions more will join their ranks in coming years. This aspect of the pandemic is most dramatically witnessed in the tiny nation of Swaziland where, within the next five years, orphans may make up more than a quarter of all children in the country. Already, many children, sometimes as young as ten, are acting as breadwinners and care-givers for households of their siblings. As the pandemic progresses, more communities and nations will be forced to redefine the basic customs and structures that have anchored them for centuries.

This economic and social upheaval is threatening the stability of nations and regions. The mounting desperation generated by the pandemic fuels lawlessness and conflict. At the same time, the disease is hindering the capacity of countries to maintain their security. Infection rates among military and police forces often exceed those in the general population. Roughly one-quarter of South Africa's army, perhaps the most important for the stability of the region and continent, is infected, while police in Mozambique are dying from the disease faster than they can be replaced. With the additional burden of a full-blown HIV/AIDS epidemic, countries already plagued by conflict and social unrest are at risk of joining the ranks of the world's failed states, in turn fuelling other regional and global conflicts.

And this is just the beginning. A sobering modelling exercise conducted by the Joint UN Program on HIV/AIDS (UNAIDS) charted the course of the pandemic in Africa over the next twenty years according to different scenarios of human response. Even in the most optimistic case, which assumes successful prevention and treatment intervention far beyond anything achieved to date, an additional 45 million people will die and more than 50 million will be newly infected. This is just Africa. If current trends continue, countries in Asia, Eastern Europe and elsewhere may soon experience the widespread economic and social shocks, to say nothing of human loss, that have thus far been largely limited to Africa. The corresponding impact on the global economy would be massive (picture here an AIDS-related 2-4 % reduction in the GDP growth of China).

That is not to say that we should concede defeat. The difference between the worst- (an extrapolation of current trends) and best-case scenarios in the UNAIDS exercise is 60 million African lives saved. Emerging epidemics in other regions could still be contained with a sufficiently vigorous response at both the local and global level. This response has begun to emerge. Leaders in China, Rwanda, South Africa and elsewhere have awoken to the threat the pandemic poses to their countries and have begun to speak frankly about the difficult steps needed to contain it. And in 2002, world leaders came together to create a new engine to drive the global response – the Global Fund to Fight AIDS, Tuberculosis and Malaria. The Global Fund was designed to be big and fast, channelling billions of dollars each year to support local programs fighting the diseases in developing countries around the world. In just five years, it has made huge strides in realising that mission, with nearly $8 billion committed to programs in 136 countries. These investments have already resulted in, *inter alia*, life-extending antiretroviral treatment for an additional 1 million people living with AIDS, a massive increase over previous levels in the developing world.

These are strong first steps, but much more commitment is required from both the rich and poor worlds if we hope to significantly alter the course of the pandemic. The Global Fund was created with the intention that it would fill a significant portion of the more than $17 billion that is now needed annually to combat the disease. Today, it is meeting just a fraction of that ambition. While the leaders of wealthy nations must dig deeper into their wallets, their counterparts in the developing world must bolster their own leadership. Initial successes by Thailand and Uganda in containing the disease highlighted the importance and potential of strong national commitment, but have yet to be thoroughly replicated in other nations.

The coming pandemic: influenza

If the world was caught off guard by HIV/AIDS, it has been given ample warning with influenza. Experts on the virus have been saying for years

FIGURE 5.3   The world is overdue for a severe influenza pandemic.

that the world is overdue for an influenza pandemic (Figure 5.3). The avian strain of the virus that made the headlines throughout the winter of 2005–06, dubbed H5N1 after its two defining proteins, first appeared nearly a decade ago in Hong Kong, prompting the slaughter of 1.5 million chickens in the province before withdrawing to a biological safe house within aquatic birds. It re-emerged again in China in 2003 and has since spread across East Asia, and recently to Europe and Africa, claiming the lives of millions of poultry and the occasional human. Even if this particular strain never develops the characteristics necessary to cause a human pandemic, it is only a matter of time before such a strain does emerge.

Humans are also no strangers to the destructive potential of the influenza virus. Over the past three centuries, there have been ten major flu pandemics, the most devastating of which occurred in 1918–19. In less than two years, this 'Spanish Flu' sickened more than 500 million people around the globe and, according to recently revised estimates, killed 50–100 million. In contrast to the endemic strains of the virus that now annually threaten the young and elderly, the majority of people who succumbed to this pandemic were aged between twenty and thirty-five. Otherwise healthy young adults would develop a cough and die within a few days. The particularly deadly nature of this strain of the virus was reaffirmed when the US Centers for Disease Control recovered a sample of the virus from a frozen victim in Alaska. Laboratory mice infected with the 1918 strain had 39 000 times as many virus particles in their lungs as those with a more modern strain.

Two other flu pandemics have since swept the globe, but it is the 1918 virus that has been at the fore of experts' minds as the H5N1 strain has progressed. The high mortality rate of the new strain that has been witnessed in East Asia (roughly 50 % of those known to be infected, though it is possible that many cases go unreported) far exceeds the less than 1 % mortality caused by most flu strains since 1919. As with the Spanish Flu, young adults are the primary victims and no version of the H5N1 strain has circulated among humans for more than a century, leaving the entire global population unprotected by natural immunity.

As of late 2006, the H5N1 strain is still an avian disease, able occasionally to jump from an infected animal to a human, but not yet capable of spreading between humans (phase 3 of 6 on the World Health Organization's pandemic alert system). However, there are ominous signs that it may be developing that capacity. Influenza viruses are highly adaptive, constantly changing and exchanging their genetic material and thus their virulence and infective ability. A virus that begins in birds may manage to infect another animal such as a pig or even a human. Once there, it can 'reassort', absorbing the material necessary for human transmission from an already adapted strain of the virus. Or, as was probably the case with the 1918 flu, it can jump straight from a bird to a human and over time adapt to the conditions of its new host. Since it first emerged in Hong Kong, the H5N1 virus has reassorted more than twenty times and evidence from the human outbreaks in Turkey during the winter of 2005–06 indicates that the virus has undergone an additional three mutations in its protein sequence. Two of these are believed to be adaptations towards human-to-human transmission. In addition, the human cases in Turkey displayed lower mortality rates and greater concentration among families than those previously – another indication that the virus may be evolving towards easier human spread.

Even with reduced mortality, a human pandemic of H5N1 could cause unprecedented devastation. Once a fully adapted strain of the virus emerged, it would be nearly impossible to contain. Spread through the air and probably highly infective, it would easily travel by airline routes to every corner of the globe in a few days or weeks. Quarantines, passenger screenings and other desperation measures by panicked governments would be ineffectual, hindering the economy more than the virus. Epidemiological models constructed by researchers at the Harvard School of Public Health showed that quarantines were effective in controlling the SARS epidemic due, in part, to the eight-day lag between an initial infection and subsequent spread to others (the serial interval). In contrast, the serial interval of the 1918 flu was four days, with individuals probably being infectious before they even knew they were ill.

While difficult to predict, the human toll would undoubtedly be large. Already, 1.5 million people die each year from annual flu strains with mortality rates of just 0.1%. Conservative estimates by the World Health Organization (WHO) predict that 2-7 million people would die during an H5N1 pandemic. An extrapolation of the 1918 pandemic provides a more sobering benchmark – at current population levels, the earlier pandemic would cause 180–360 million deaths around the globe.

The effect on the global economy could be equally staggering. While the economic impact of HIV/AIDS is playing out over decades and generations, the effects of a flu pandemic would be immediate. Direct costs from illness and death and the corresponding healthcare response would be sizeable. But much more significant would be the effects of the resulting panic and disruption of global trade. The outbreak of SARS in 2003 provides a useful reference here. Though the 8000 cases of the disease had no impact on direct output, the virtual halt of travel to the region resulted in a 2% reduction in the quarterly GDP growth of the East Asia region. Simply extrapolating this to the global scale, as the World Bank has recently done, would mean an $800 billion loss to annual global GDP. In addition, a study by Citigroup projected that direct labour costs could result in a 2-3% loss of GDP. The combined effect of a roughly 5% reduction in GDP, or $2 trillion annually, is in line with independent estimates by the Congressional Budget Office of the impact of a severe (2.5% mortality rate) influenza epidemic on the US economy.

A full-blown flu pandemic could quickly dwarf the losses experienced with SARS. Deaths would number in the millions rather than the hundreds, and economic disruption would reach a similar scale. Schools would close and employees would stay at home for fear of catching the disease. Governments would shut their borders and hoard essential commodities, bringing global trade to a standstill. Those businesses that did remain open would find it difficult to operate without access to their suppliers and consumers in other countries. In a testimony before the US Congress in 2005, Dr Michael Osterholm asserted that the aspect of pandemic preparation requiring the greatest additional attention from governments and companies was this 'critical product continuity'. As support, he cited the example of the pharmaceutical industry. More than 80% of the raw materials used by US pharmaceutical manufacturers are produced outside the country and, with the near-instant global shipping now available, only thirty days of extra material are usually kept in reserve. Already in the midst of one public health crisis, nations would face another as a wide range of pharmaceutical products would become unavailable. Producers of other essential goods would face similar challenges. Taking into account some of these effects, the Lowy Institute, an Australian think-tank, has modelled the impact of an 'ultra' pandemic – similar to the 1918 pandemic but with higher mortality among the elderly – at a 12.6%, or $4.4 trillion, reduction in annual global GDP.

Science does not yet offer any easy solutions to this threat. Influenza vaccines are still produced through a labourious process, first developed in the 1950s, whereby a sample of the virus is slowly grown on live chicken eggs. As a result, it would be nearly six months from the first emergence of a human influenza until an effective vaccine was available. Efforts are under way to prepare vaccines from the H5N1 strains currently circulating, but given the frequent mutations of the virus, these would probably have limited efficacy in the event of a pandemic. New methods such as reverse genetics and cell-culture technology are still in development. Even once a vaccine is developed, it will only be available to a fraction of those at risk. With current production capacity, vaccine manufacturers would be able to produce a maximum of 750 million doses in the first year of a pandemic – enough to immunise 12% of the world's population. These manufacturers are located in just nine countries and once a pandemic begins, it is likely that the governments of these privileged few will prevent exports of vaccines to ensure sufficient supply for their citizens. Medications, such as the antiviral oseltamivir ('Tamiflu') being stockpiled in 2006 around the world, will also play an important role in a pandemic response as treatments and prophylaxes, but these suffer from similar production limitations as vaccines and it is unclear how effective they will be against a new strain of the virus. In fact, a study published in 2006 in the *Lancet* by Jefferson and associates found that oseltamivir had no effect on current avian influenza strains and could even facilitate the spread of the virus.

In sharp contrast to the HIV/AIDS experience, the global community has rapidly mobilised in preparation for a flu pandemic. Strategies for preventing, containing and mitigating a pandemic have been developed by the WHO and other international bodies and, in an event virtually unheard of in global health, donors pledged significantly more than the estimated $1.2 billion required to help poor nations implement these plans at a fund-raising conference in Beijing in January 2006. Funding is also flowing to support efforts to overcome the technological and capacity constraints to vaccine and treatment production. To be sure, these are only first steps, which would have small effect were a human flu pandemic to emerge tomorrow. But they far surpass anything witnessed in the first two decades of the HIV/AIDS pandemic. The causes for this striking divergence range from the less charitable (HIV/AIDS is killing millions of 'them' while avian flu will kill millions of 'us') to the more charitable (we have learnt lessons from HIV/AIDS and appreciate that new viruses threaten the entire global community).

A 'wise and frugal' response

If we now accept that pandemics should be among the top priorities for global action, the question then arises of what that action should be. The global response to HIV/AIDS was grossly deficient for two decades and even recent commitments are still insufficient to slow and halt the expanding pandemic. While stronger, the actions now being taken to prevent and mitigate a flu pandemic face significant limitations. What then should our response look like? Here again it will be useful to envision a global government and the mechanisms and systems it would put in place to address the challenges of pandemics. This, in turn, will provide us with additional insight on the strengths and weaknesses of our current global structures.

Firstly, let us determine the key elements of an effective pandemic response that must be provided at the global level. Earlier in this chapter we established that global public goods are prone to a high degree of market failure and as such require intervention by the global community or, in the case of our exercise, global government, to ensure their provision. For pandemic control, there are four primary areas where such intervention is necessary: communication and co-ordination; effective distribution of resources; ensuring best possible science and technology; and intervention for rapid containment of emerging epidemics.

Communication and co-ordination

Rapid and transparent flow of information is essential to containing and mitigating a pandemic, particularly in the case of a virulent and mutable disease such as influenza. Mutation of the H5N1 virus, such as that detected in Turkey in January 2006, can have immediate and profound implications for the efforts of governments, vaccine researchers, businesses and many others around the world. Yet left to the market, it is doubtful that this critical data would reach all who need it with the speed and accuracy required. Co-ordination is equally important. Each individual nation could conduct the same studies on the implications of those mutations, but it would be a highly inefficient use of limited resources.

This is the least complex of the four areas, both in theory and practice. A global government would presumably form a central agency to conduct this function. Countries would be required to immediately report critical information about an emerging epidemic to this agency, which would assist in organising and coordinating a global response. In the current global system, there is an institution that performs this role relatively well – the World Health Organization. WHO's ability to fulfil this function was recently enhanced when the World Health Assembly, the council of nations which serves as WHO's Board, agreed to new International Health Regulations. First developed in 1951, this is the primary international treaty governing nations' actions to prevent and contain threats to global health. By 2000, it was clear that the regulations required updating: only three diseases with relevance today – cholera, plague and yellow fever – were treated in the original. In addition to including a more comprehensive set of modern infections, the new regulations, which will come into force in 2007, require all nations to report 'public health emergencies of international concern' (i.e. outbreaks of major infectious diseases) to WHO. The central challenge now is one of enforcement – if China does not want to report honestly an outbreak, as with SARS in 2003, there is little that WHO can do about it.

Effective resource distribution

Containment of a pandemic is largely reliant on the ability of local systems to detect, report and respond to a new outbreak of a disease. That ability varies greatly among nations according, principally, to their wealth. A human flu pandemic would take a much different course were it to first emerge in Chad rather than Canada. Once a pandemic has gained a foothold, the assistance required by poor nations to mitigate its effects rises dramatically. But, as the experience of HIV/AIDS has clearly demonstrated, the costs of inaction far outweigh those of even the most expensive interventions.

To inform our thinking on this function, it is useful to consider the experience of the United States. In many ways, the United States is a microcosm of a world with a global government, with individual sovereign states each of which cede authority to the federal government to manage issues that either unite (national security) or divide (inter-state commerce) them. Citizens pay both state and federal taxes to support the provision of public goods by each respective government. So too would our global government levy taxes to support the provision of global public goods, including pandemic control. Ideally, these global taxes would exactly meet the level of expenditure required to ensure those goods, rising and falling according to changing needs (ignoring, for the sake of this exercise, Keynes and a host of macroeconomic complexities). In allocating this funding, the global government would make decisions based solely on where it was most needed and could be used most effectively.

The current system is far from this ideal. Contributions to global pandemic control are, for the most part, voluntary and are determined largely by the political exigencies within and between nations. As a result, global efforts to control HIV/AIDS and other pandemics are facing a serious shortfall, while the threat of an influenza pandemic has received contributions in excess of estimated needs. The greatest resemblance to a global tax is the system of United Nations dues, though as the United States demonstrated when it withheld payments for years, there is no taxman to ensure their collection. Where and how funding is spent is also frequently dictated by political realities. For example, Australia prefers to focus its funding in its immediate region, while the United States will not fund groups that support abortion.

There have been several recent positive steps in this area. The governments of Brazil, Chile, France, Norway and the United Kingdom have begun to implement a new levy on international airline travel to bolster global efforts to control HIV/AIDS, tuberculosis and malaria. The UK government proposed and is now piloting an International Finance Facility which frontloads aid commitments for problems, notably diseases, which are less expensive if addressed earlier – not a tax as such, but a more rational method of distributing resources. And the Global Fund was created in 2002 to rise above politics in the allocation of resources for HIV/AIDS, tuberculosis and malaria. All of the more than $5 billion it has committed as of mid 2006 has been solely on the basis of need and technical merit – a fact that is clearly reflected in the distribution of its funding across 132 countries and to the full range of proven health interventions. While global levies, such as the air-travel initiative, will always be a hard sell in the current system, the early success of the Global Fund suggests that the global community should embrace and adopt this model more broadly.

Scientific and technological resources

Medical technology is, of course, the most critical component of pandemic control. The development of a cure for HIV/AIDS or a universal influenza vaccine would dramatically alter our ability to control these diseases. But this area suffers from one of the most pronounced examples of market failure. Currently, only 10% of all research and development expenditures are devoted to diseases that affect 90 % of the world's population. For private companies, the choice is clear – profits lie in the non-communicable and cosmetic afflictions of the rich world, not the deadly plagues of the poor one. Even when products exist, firms are often unwilling to produce the necessary amount due to the lack of a clear and predictable market. Other disincentives, including high liability and unknown time horizons, have deterred investment even in influenza and other infections that also threaten wealthy nations.

As both Adam Smith and Thomas Jefferson would agree, our global government would not simply provide this function through a massive centralised research agency. Rather it would ensure that the necessary incentives were in place for the private sector to invest its own resources. These could include tax breaks for research in infectious diseases, advanced purchasing of vaccines or other high priority medical interventions, and liability protection. More simply, our world government could aggregate global expenditures on certain diseases and products, ensuring that private firms not only invest in research and development for these diseases, but that they also increase production capacity to meet the global demand for these essential goods. Lastly, this government would have a single regulatory standard so that, once approved, new medical products would be available for use in all countries.

In recent years, numerous initiatives have been launched to attempt to overcome this paucity of research and production, driven primarily by the Bill and Melinda Gates Foundation. A series of new organisations, such as the Malaria Vaccine Initiative and the Institute for One World Health, were created to leverage the resources and expertise of private researchers to address critical gaps in the fight against infectious diseases. The Global Alliance for Vaccines and Immunization has created a central pool of resources to ensure the supply and delivery of vaccines for a number of prevalent infections. Individual governments have attempted to put in place incentives to address their own vulnerabilities to infectious diseases and bioterrorism. Project Bioshield in the United States, for example, provides guarantees of government purchase and regulatory fast-tracking for new products to control certain diseases. And, in July 2006, the G8 made an initial commitment to launch a multi-billion-dollar advance purchase mechanism to stimulate vaccine development. But there are still significant gaps and promising opportunities have gone largely unrealised. The billions of dollars channelled through the Global Fund could be better leveraged to increase research into and production of much needed tools to combat HIV/AIDS, tuberculosis and malaria. These and other innovative solutions should be seriously discussed and considered by the global community or we risk fighting these diseases with the same antiquated medications and tools decades from now.

Global intervention capacity

Whether through inability, negligence or deliberate obstruction, many nations will not reveal or effectively respond to an outbreak of disease within their borders. The likelihood and implications of such inaction were highlighted by the many months of obfuscation of the SARS outbreak by the Chinese government, which enabled the disease to spread around the world with corresponding global economic consequences. More recently, the governments of Nigeria and Turkey initially denied outbreaks of H5N1, and there is evidence that China may still be concealing the magnitude of the disease within its borders. The emergence of a human flu pandemic in Myanmar or North Korea would surely be met with even greater obstruction. In these cases, the only recourse for the global community to avert a potentially devastating pandemic is to rapidly insert itself to implement the necessary measures at the local level.

Here again, the experience of the United States is a useful reference. Health is primarily a state responsibility. But in 1946, the federal government transformed the Office of Malaria Control in War Areas into a new central disease-fighting agency, the Communicable Disease Center (CDC). Over the coming decades, this new agency was the focal point for the nation's efforts to control infectious diseases, acting as a central reservoir of information and expertise to coordinate and enhance the work of individual states. The change of the agency's name in 1970 to the Center for Disease Control (and a decade later to 'Centers' after a bureaucratic reshuffling), reflected the shifting disease burden within the country – malaria, smallpox and other infections had been eradicated, while non-communicable threats such as heart disease and cancer were rising dramatically. Nevertheless, CDC retains its original mandate and would be the nation's chief tool in the event of a human flu pandemic.

Contrary to what one might assume, CDC does not have legal authority to intervene directly in response to a local outbreak without the invitation of the state. In fact, states are not even legally bound to report new cases of infectious diseases to CDC. In practice, however, decades of close collaboration and exchange of personnel mean that a state would never refuse CDC's assistance in the event of a disease crisis. Our global government would presumably adopt a similar model: a CDC of the world with either informal power of intervention, as in the US model, or more direct legal authority.

This function poses a much more imposing challenge for the current global system. Unlike US states, nations have a strong interest in not allowing an outside body to cross their borders at its discretion. The WHO is currently the closest structure to a global CDC. However, despite recent revision and years of negotiation, the International Health Regulations make no mention of authority of intervention. Even if the International Health Regulations did grant this authority, the WHO would have little or no power of enforcement if a country is uncooperative.

Where then does that leave us? We could create a new international body with more authority than WHO. But what would grant this agency any more power to enforce that authority? We could strengthen the WHO through renegotiation of the International Health Regulations in spite of strong resistance and the limitations of enforcement. We could create a Security Council for health, which would review cases of nations endangering the world's health and authorise necessary action by the global community, though it is unlikely such a body would be able to operate with the speed required to contain an emerging pandemic. Or should we abandon this core function altogether and rely on the diplomatic tools currently available to us while preparing to mitigate pandemics that arise in situations we cannot control? Current trends favour a coalition of the willing – maximum collaboration between like-minded states for mutual self-interest. The group of Australia, Canada, China, France, Indonesia, the United Kingdom, the United States and others, which met in Ottawa in 2005 to discuss joint efforts to combat an avian flu pandemic, exemplifies this approach.

Human response

Much attention has been given to events that have supposedly dramatically altered the course of human history over the past hundred years or so. Countless articles and books have been written about the effects of the end of the Cold War or the rise of the internet on the interactions of nations and individuals, and the phrase 'since September 11' is now seemingly requisite for all commentary or analysis. But though it has arguably had a no less dramatic impact, the changing nature of disease has gone largely unremarked. A child born in the United Kingdom today can expect to live roughly thirty more years than if she had been born a century ago – a boon largely attributable to her freedom from most of the viral and bacterial threats that plagued her ancestors. At the same time, a child born in Botswana today can expect to live thirty years less than if HIV/AIDS had not engulfed her country. These trends have as profound, if not always as readily visible, an impact on human society as any political or cultural development. As US Supreme Court Justice Stephen Breyer summarised well:

> We think of our current era as 'the nuclear age' or 'the information age,' but I am increasingly convinced that when we look back at this period of history from years ahead, we will see that this was the 'age of AIDS'.

The formative power of events such as the of the Cold War or the September 11th terrorist attacks lies not in the events themselves, but in the human reactions to the challenges and opportunities they generate. So it is with pandemics. How we collectively chose to respond to the next pandemic (to say nothing of the HIV/AIDS pandemic currently upon us) could mean the difference between hundreds and hundreds of millions lives lost. As we outlined above, the nature of pandemics has changed dramatically over the past few decades. The question now is whether we will adapt to meet the new challenges they pose. To reiterate Charles Darwin's famous remark:

> It is not the strongest of the species that survive, nor the most intelligent, but the one most responsive to change.

FURTHER READING

Garrett, L. (2005). 'The next pandemic?', *Foreign Affairs* **84**, 3–23.

Osterholm, M. T. (2005). 'Preparing for the next pandemic', *Foreign Affairs* **84**, 24–37.

Shilts, R. (2000). *And the Band Played On: Politics, People, and the AIDS Epidemic*. New York: St Martin's Press.

Taubenberger, J. K. and Morens, D. M. (2006). '1918 influenza: the mother of all pandemics', *Emerging Infectious Diseases* **12**, 15–22.

UNAIDS (2005). *AIDS in Africa: Three Scenarios to 2025*. Geneva: Joint United Nations Programme on HIV/AIDS.

6 Surviving natural disasters

JAMES JACKSON

Introduction

Natural disasters pose a rather different threat to that of disease. Volcanoes, tsunamis, hurricanes, floods and earthquakes are rare events, but when they hit they have the potential to obliterate communities in seconds. In this chapter I will focus on the threat posed by earthquakes and will discuss in particular the dreadful vulnerability of megacities in the developing world.

Surviving earthquakes: vulnerability in the modern world

It has already been a bad century for earthquakes, with well over a third of a million people killed in just four catastrophes: at Bhuj in India (2001; 20 000 dead), Bam in Iran (2003; 40 000 dead), Bandah Acheh in Indonesia (2004; over 200 000 dead) and Muzaffrabad in Pakistan (2005; 80 000 dead). As a seismologist, I am frequently asked whether disastrous earthquakes are more common now than they were in the past. It is an easy question to answer. Over the last thousand years, I know of 110 earthquakes that each killed more than 10 000 people (Figure 6.1). Yet 34 of them happened in the last century. In other words, a disaster of that magnitude happened, on average, about every 11 years from 1000 to 1900, about every 3 years for the next century, and now virtually every year.

Survival, edited by Emily Shuckburgh. Published by Cambridge University Press.
© Darwin College 2008.

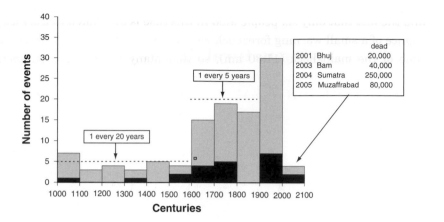

FIGURE 6.1 Histogram of the number of earthquakes killing more than 10 000 (gray) or 50 000 (black) people per century over the last 1000 years. There are 113 earthquakes in this figure, with 34 of them occurring in the last 100 years alone. Until about 1600 they averaged about one every 20 years, increasing to nearly one every 5 years between 1600 and 1900. This century we have had nearly one every year.

The answer to the question is obviously 'yes', and in fact the dramatic increase in such disastrous earthquakes has been since about 1600, but the emphasis is on *disastrous* – the effect is not caused by any change in the natural behaviour of our planet, as the earthquakes themselves are no more or less frequent now than they ever were, but is related to the way we now live, to our historical response to the natural environment, and to the rapidly increasing global population. A closer look at these recent disasters reveals the situation in which we now live. We will start with an apparently innocuous example.

A bull's-eye hit in the desert

In February 1994 the small desert village of Sefidabeh in south-east Iran was destroyed by an earthquake of moderate size (magnitude 6.1). Most of the 300 or so buildings in Sefidabeh collapsed, having been built of adobe, or sun-dried mud-brick, the traditional indigenous building material for desert villages, from which both walls and heavy roof domes are constructed. Adobe is a notoriously dangerous material in earthquakes, and the fact that only six people died in this case is attributed to the lucky chance of a small warning foreshock 24 hours beforehand, and to the local time of the mainshock (11.30 am), so that many people were outdoors anyway. But there is more to this story.

Sefidabeh is a desperately remote and inhospitable location, sandwiched between the two deserts of the Dasht-e-Margo (*lit.* 'desert of death') in Afghanistan and the Dasht-e-Lut (*lit.* 'barren desert') of south-east Iran; one of the very few stops on a long, lonely trans-desert trade route between north-east Iran and the Indian Ocean. It is the only habitation of any size for nearly 100 km in any direction, and yet the earthquake apparently targeted it precisely. Was this a case of extreme bad luck, or is there more to it?

Earthquakes happen when faults move. Faults are giant, planar knife-cuts through rock, in the case of Sefidabeh extending 20 km along the surface and 10 km deep into the Earth. The two sides are held together by friction, but occasionally jerk past each other in earthquakes, shaking and vibrating as they do so. The vibrations travel round the Earth as sound waves and are recorded, as seismograms, by permanent worldwide networks of sensitive instruments.

Our modern technology and understanding of earthquake-related faulting give us a forensic-like ability to work out exactly what happened at Sefidabeh in 1994. Analysis of the earthquake seismograms shows that the fault which moved was inclined at about 45° to the horizontal, and its movement pushed one side towards, and on top of, another. More detailed information comes from analysis of space-based radar measurements, which determine precisely the changes to Earth's surface that result from the fault motion, and allow us to fix exactly the location of the fault (a few kilometres south of Sefidabeh), its length (approximately 12 km), how deep it goes (approximately 10 km) and how much it moved (approximately 2 metres). In particular, the radar analysis shows that the top of the fault was about 4 km below the ground surface; in other words, that the slip on the fault was entirely confined below ground and did not reach the surface at all.

Instead of a fault rupture or scarp at the surface, what formed was a fold. As a useful analogy, imagine sliding the top half of a telephone directory over the bottom half towards the binding: the slip surface would be the fault but, because of the binding, a fold develops at the end of the fault. A fault of this type, in which slip fails to reach the surface, is called a 'blind' fault, and is shown in Figure 6.2a. Slip in a single earthquake is only a metre or two, but repeated earthquakes over hundreds of thousands of years cause the fold to grow into a ridge. The ridge adjacent to Sefidabeh is about 100 metres high and a dominant feature of the landscape. Furthermore, the landscape itself preserves clear evidence that the ridge is young in origin and had been growing as a result of previous earthquakes (see Figure 6.2b). Sefidabeh is built on an old 'alluvial fan', formed where an ephemeral river that used to flow through the ridge discharged its water, and with it sediment known as 'alluvium', onto the desert plain. But as repeated earthquakes caused the ridge to grow, not just in height but also by increasing its length towards the north-west, the river had to incise a gorge through it, eventually becoming blocked, and forming a lake. Finally the river abandoned this course altogether, and now flows round the north-western tip of the ridge instead. The old lake beds remain, now dry and elevated 70 metres above the desert plain. From the age of the sediments within them we can date the switch in the river course to about 100 000 years ago.

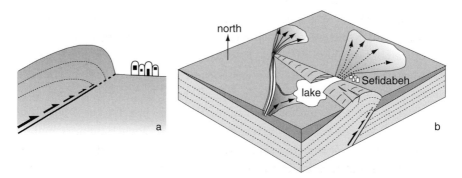

FIGURE 6.2 (a) Schematic cross-section of a 'blind' fault, adjacent to a village (not to scale). Slip on the fault (arrows) dies out towards the surface, which deforms by creating a fold. (b) Simplified cartoon of a blind fault and its fold at Sefidabeh in eastern Iran. The fold is about 10 km long. As the fold grew in repeated earthquakes, a river that used to flow across it first incised to make a gorge, then flooded to make a lake, and was finally abandoned when the river course switched to flow round the north-west end of the fold.

Thus, well before the earthquake, all the signals were there in the landscape that Sefidabeh was in a vulnerable location – if people had only seen them and known how to read them (they hadn't: Sefidabeh was too remote for anyone to have noticed). 'Blind' faults of this type are very common in Iran, as the whole country is being squashed in a north–south direction between the converging Arabian and Eurasian plates at about 25 mm per year. But our ability to recognise the folds created at the surface by blind faults like the one at Sefidabeh dates only from 1980, when one moved in the El Asnam earthquake in Algeria. Earlier devastating Iranian earthquakes of modern times that occurred on similar blind faults include those at Ferdows in 1968 (magnitude 6.3, approximately 1000 killed) and Tabas in 1978 (magnitude 7.3, approximately 20 000 killed). In neither of these cases were the causative faults recognised at the time, though, in retrospect, they are clear in the landscape. The city of Bam, destroyed in 2003 (magnitude 6.8, approximately 40 000 killed) was also located on a blind fault of this type, which was recognised beforehand, though the faulting in that earthquake was more complicated. In each of these cases, at Sefidabeh, Ferdows, Tabas and Bam, the earthquakes involved blind faults that we now know about in detail. But they share another characteristic too; each place was the only substantial habitation for tens of kilometres in any direction, and yet the earthquakes apparently targeted them precisely. We have not yet explained why the earthquakes apparently scored bull's-eye hits in each case.

Faults and water

The answer is water. Sefidabeh means 'white water', and the village obtains its water from the white lake beds in the uplifted ridge, which leak in a series of springs at its base. The fault is responsible for the sub-surface aquifer of the lake beds, and ensures their continual uplift and elevation above the plain, causing the formation of springs. Sefidabeh is the only place where it is possible to live and attempt a meagre agricultural existence in this extremely inhospitable desert environment, as it is the only place with water. It is the fault that provides the water, but the fault may kill you when it moves.

This situation is common in Iran. The country is mountainous except for flat regions in the interior, which are barren salt flats. The mountains provide the water, so habitations are common at the foot of range fronts, many of which exist because they are elevated by movement on faults, just like the ridge at Sefidabeh. For centuries, the indigenous people have exploited this situation. Some horticulture is possible on the toes of alluvial fans coming off the ranges, on the finer-grained material away from the coarse debris adjacent to the steep slopes; but only if water is available. The water table is usually elevated at the range front, sometimes exaggeratedly so if there is an active fault, because repeated grinding of rocks on the fault creates a very fine, impermeable clay (called 'fault gouge') that can act as an underground dam to the water table, elevating it still further. Tunnels are dug, by hand, back to the range through the semi-consolidated fan material, to tap the elevated water table at the range front. In Iran, these tunnels are called *qanats*, and are one of the glories of the ancient Persian civilisation. They can be several tens of kilometres long, up to 100 metres deep at the range front, and are marked at the surface by lines of circular craters, where vertical shafts are sunk down to the tunnels to provide access, ventilation and removal of excavated material during construction. Many have been in continuous use for centuries, and the oldest are thought to have been dug more than 2000 years ago. More details of qanat history and construction are given by in a book by Anthony Smith and in an article by Hans Wulff.

Qanats provide fresh, continuous supplies of water, with little evaporitic loss, to thousands of villages in the deserts of Iran and other countries in the Middle East and central Asia. They are engineering wonders, and the very lifelines by which existence is possible for many. For example, the oasis city of Bam, destroyed in 2003, uses water from qanats that tap the nearby aquifer above a blind fault to feed the date-growing region of Baravat, one of the most famous date-producing regions in the Middle East (Figure 6.3). It is this fault-controlled water supply that determines where Bam–Baravat is located, and why it was destroyed. Elsewhere in the desert, qanats bring water to fantasy pleasure-gardens, with cascading pools of water surrounded by trees and pavilions, such as at the central Iranian town of Mahan (Figure 6.4a) and the oasis of Tabas (destroyed in the 1978 earthquake).

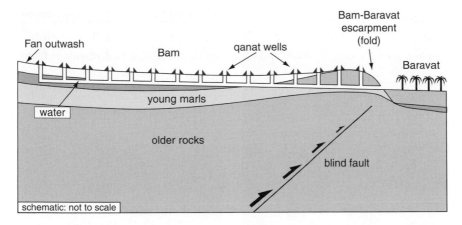

FIGURE 6.3 Schematic section of the irrigation tunnel ('qanat') systems feeding the date growing region of Baravat adjacent to Bam. The blind fault has created a fold which ponds the sub-surface water flow through the young outwash from the adjacent mountains, as the underlying lake beds (marls) are relatively impermeable.

Living in the desert

For centuries, Iranian civilisation and desert existence has lived with, and exploited, the link between mountains, faulting and water supply. This relationship can be illustrated by three short examples.

The beautiful desert oasis of Tabas, properly known as Tabas-e-Golshan (*lit.* 'Tabas the flower garden'), visited by Marco Polo in the thirteenth century, was destroyed in 1978 by movement on a series of blind faults, whose folds are clearly visible in the landscape. The water supply for Tabas comes from the adjacent range front by qanats which penetrate the nearby fold. The fold is also cut by the ephemeral Sardar river, which has incised a deep gorge through the rising ridge, in the same way as the stream at Sefidabeh, before its gorge was finally abandoned (Figure 6.2b). But unlike at Sefidabeh, the Sardar river at Tabas still maintains a gorge through the rising fold, though it only occasionally contains water. In the past, this river caused problems of its own, when flash summer thunderstorms in the mountains produced great volumes of water that were trapped within the Sardar gorge and then discharged when the river emerged through the fold, to flood Tabas. The local response, attributed to Shah Abbas (seventeenth century), was to build a dam or water-gate where

FIGURE 6.4 Living with earthquakes and faults. (a) The water gardens at Mahan, in the desert between Bam and Kerman, fed by qanats from mountains several kilometres away. (b) The seventeenth-century water-gate across the Sardar river at its narrowest point upstream from the oasis of Tabas-e-Golshan. (c) View over the mud roofs of a village near Ferdows in the desert of eastern Iran. The blue roofs are coated with a clay which is impermeable fault gouge excavated from a quarry at the foot of the range front in the background, where the fault comes to the surface.

the river leaves the mountains (Figure 6.4b); but one in which there is a vaulted arch at the base to allow the bed-load of the river through, while limiting the head of water to a height that was manageable. This ingenious, maintenance-free solution to flood control has stood for 350 years, and is still effective today.

After the devastating 1968 earthquake at Dasht-e-Bayaz in eastern Iran (magnitude 7.1, approximately 12 000 killed) several qanats were cut and offset by horizontal movement on the causative fault, which slipped up to 4 metres in places. These qanats were then either abandoned or repaired by reconnecting the offset channels. There is clear evidence on the ground, and in aerial photos (Figure 6.5) for earlier generations of qanats that had been offset and abandoned in previous earthquakes. Even more remarkable are subsidiary, minor, qanat tunnels that had been dug long ago so as to feed into the main channels, and which followed precisely the line of the 1968 fault rupture. These side-tunnels exploit a change in water-table level across the fault, caused by the impermeable clay fault gouge, to tap and increase the water flow into the main tunnels. Thus the local tunnel-builders were aware, and had exploited, the fault-related hydrology for a considerable time before the modern earthquake (and before seismologists or geologists understood any of this).

Figure 6.4c shows a view over a desert village near the town of Ferdows (destroyed in 1968, 1000 killed), towards a fault-bounded range front. The houses are built of adobe walls, with roofs either of mud-brick domes or of poplar logs laid horizontally and sealed with mud. Most of the roofs have a distinctive blueish hue, caused by the clay that is used to seal against the winter rains. The clay comes from a quarry at the foot of the range front, and is in fact the fault gouge itself, made from finely ground volcanic rocks. The material is suitable for this purpose, being fine and relatively impermeable. Such blueish mud roofs are a common sight at range-front villages in eastern Iran.

Villages, mega-cities and population growth

The point of the examples cited so far is to illustrate that, for centuries, desert-rim existence in Iran had established a way of living with earthquakes. Earthquake faulting, and the topography it produces,

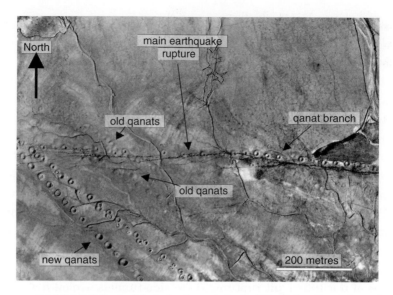

FIGURE 6.5 Aerial photo of qanats in the Nimbluk valley, taken after the 1968 Dasht-e-Bayaz earthquake in eastern Iran. The earthquake fault rupture runs east–west across the centre of the picture and moved horizontally, with the north side sliding to the west. Multiple generations of qanats are visible, the most recent were cut by the 1968 faulting, but these were replacements for earlier qanats, whose lines of craters are now heavily eroded, that were presumably abandoned after earlier earthquakes. One qanat follows precisely the line of the fault rupture, increasing the water flow into the main north-west–south-east qanats by tapping an underground change in water level, ponded by impermeable clay gouge on the fault. (Photo courtesy of N. Ambraseys.)

is largely responsible for the water resources and for the locations of habitations and agriculture, as well as of some building materials. Occasionally, earthquakes moved the faults, and villages were destroyed, but the repeat times of earthquakes on individual faults are likely to be measured in thousands of years and they are most unlikely to recur on a timescale relevant for human memory. When earthquakes do occur, the destruction, and particularly the mortality, can be shocking, because of the vulnerable local building styles. Thus in the town of Tabas in 1978, more than 80% of the population (11 000 out of 13 000) were killed outright; at Bam in 2003 the figure was nearer 30%. Most places are, nonetheless, rebuilt and resettled because their location is, in the end, determined by where water is available and agriculture is possible. In the past, when rural populations were relatively small and dispersed, the frequent earthquakes of magnitude 6–7 that occur in Iran would kill typically a few hundred or thousand people. A modern example is the earthquake near Zarand, north-west of Kerman, in February 2005 (magnitude 6.4), which destroyed two villages, killing 500.

The problem is that villages grow, and have grown, rapidly, while building quality remains equally vulnerable, though it may change from weak adobe houses to poorly built multi-storey apartment blocks, and so mortality rates remain appallingly high. Thus the village of Sefidabeh (6 killed in 1978) can become the large rural town of Tabas (11 000 killed in the town in 1978; 20 000 including other villages of the oasis), or the small cities of Bam (40 000 killed in 2003) or Rudbar (40 000 killed in 1990), or the mega-city of Tehran, which now has a daytime population of 10–12 million. The case of Tehran is instructive. It is situated at the base of the Alborz mountain range front (Figure 6.6), which is elevated by movement on an active fault. Several other active faults are also situated nearby. Until as recently as the 1930s, Tehran's water supply came entirely from qanats that penetrated these faults, though now the water table has been considerably lowered by groundwater extraction, and the qanats are no longer operative. In former times, the site of Tehran was occupied by relatively small towns on a major trade route. These predecessors of modern Tehran were damaged or destroyed completely in earthquakes of probable magnitude about 7 in 855, 958, 1177 and 1830, but the number of killed was probably quite small by modern standards, perhaps measured in hundreds or thousands. The modern Tehran is a mega-city that grew rapidly on the same site in the later twentieth century. While the Tehran site was occupied by relatively small towns, the city of Tabriz was always bigger, more prosperous and far more important as a trade-route crossroads. Tabriz was devastated by major earthquakes on its nearby faults in 1721 (more than 40 000 killed) and 1780 (more than 50 000 killed), at a time when the population was a small fraction of today's.

The message is clear: there is no sign that the concentration of population into large towns and cities in Iran is accompanied by a decrease in the mortality rate during earthquakes. Many major towns and cities are situated adjacent to range fronts and faults, in places that made sense

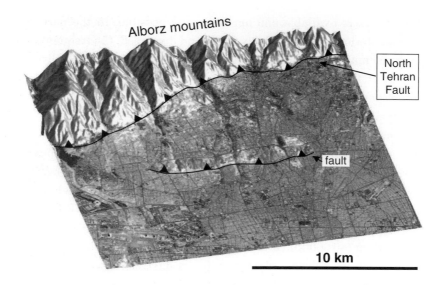

FIGURE 6.6 Perspective view looking north over Tehran, formed by drap-ing a satellite image over a digital topographic model. The North Tehran Fault runs along the foot of the Alborz mountains. Another fault forms the ridge in the centre of the city; note how rivers incise gorges upstream of the fold, as in Figure 6.2b. This ridge is the location of a new hospital in Tehran as well as of a landmark telecommunications tower. Until the 1930s, Tehran's water supply came from qanats that tapped the faults adjacent to the mountains.

when they were initiated as agricultural settlements, and they retain that vulnerability to earthquakes. In such places, earthquakes that in the past killed a few hundred or thousand people will now kill tens or hundreds of thousands, or more.

The link between how and where people live and earthquakes is partic-ularly dramatic in Iran, but for many other parts of the great earthquake and mountain belts that run from Italy to China, the situation is similar. Throughout this region the mountains are largely created by fault move-ment in earthquakes, pushing blocks of rocks on top of each other, all ultimately the result of the ongoing collision between the Eurasian plate and the African, Arabian and Indian plates to the south (Figure 6.7). Large tracts of this area are either low, barren, inhospitable deserts, or high, inaccessible and also inhospitable plateaus, such as Tibet. Habitations

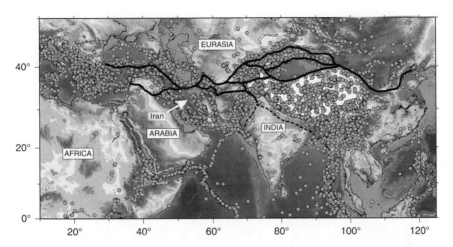

FIGURE 6.7 A map of earthquakes (gray dots) in the Mediterranean, Middle East and Asia in the period 1964–2002. The earthquakes follow, and are ultimately responsible for the growth of, the mountains that run from Italy to China, and are caused by the northward motion of Africa, Arabia and India into Eurasia. The ancient east–west trade routes, shown as black lines, follow the earthquake belts along the edges of the mountains, avoiding the flat inhospitable deserts adjoining them. Habitations along these routes have evolved from small villages, into towns, and now cities of a million or more people. Thus earthquakes which, in the past, killed a few hundreds or thousands, will now kill many more, when they recur.

concentrate around the edges of these regions, at the range fronts, because their locations are on trade routes, are of strategic importance controlling access, or are near water supplies. But the range fronts are often faults, and many of these places have been destroyed in past earthquakes.

Thus we come to the 2005 Pakistan earthquake (magnitude 7.5) which destroyed the Himalayan town of Muzaffrabad, killing at least 80 000. The fault responsible for this earthquake also pushed one side (Tibet), over another (India), and is ultimately responsible for building the Himalayan range itself. The part that moved in 2005 was a 100-km section of a much longer system of faults that stretches east–west for 3000 km, from north-west Pakistan to Assam (Figure 6.8), all of which moves because India and Tibet converge at about 20 mm per year. Figure 6.8 shows the locations of earlier earthquakes on this fault system, some of them much larger than the one in 2005. Three such events happened in the last century, in 1905,

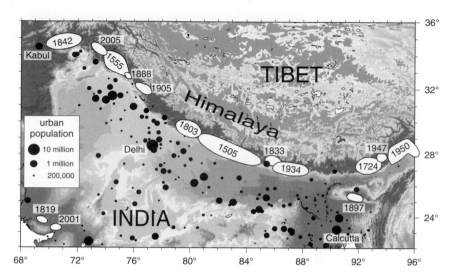

FIGURE 6.8 White areas with dates are the sites of large earthquakes along the Himalayan range front between 1500 and 2005. Also shown are significant population centres in the region, many of which are vulnerable to future repeats of these past earthquakes. (Data from Roger Bilham.)

1934 and 1950. Each one killed a few thousand people, but since those times the population of the Ganges basin has increased dramatically, and it is now one of the most densely inhabited regions on Earth, with many cities of over a million people. The flat plain of the Ganges valley contains thick sequences of water-saturated sand and mud washed off the rising Himalaya and deposited by the river. When shaken in strong earthquakes, the sediments liquefy, releasing water which spouts to the surface as springs and sand 'volcanoes'. The effect is similar to walking on a sandy beach just washed by an incoming wave: water-saturated sand 'flows' easily through one's toes. Such liquefaction of sand is well known, and is described dramatically by a geologist who observed the 1934 earthquake in Bihar:

> As the rocking ceased... water spouts, hundreds of them throwing up water and sand, were to be observed on the whole face of the country, the sand forming miniature volcanoes, whilst the water spouted out of craters, some of the spouts were quite 5 feet high. In a few minutes – as far as the eye could see – was a vast expanse of sand and water, water

and sand. The road spouted water, and wide openings were to be seen across it ahead of me, and my car sank, while the water bubbled and spat, and sucked, till my axles were covered. 'Abandon ship' was quickly obeyed, and my man and I stepped into knee deep water and sand and made for shore. (From: 'The Bihar-Nepal earthquake of 1934', *Memoirs of the Geological Survey of India*, vol. 73, p. 34, 1939.)

A large earthquake on the Himalayan front will cause such effects over a substantial part of the Ganges valley, liquefying an area perhaps 100–200 km long and a several tens of kilometres from the mountain front. The consequences of such liquefaction for multi-storey buildings is to make them sink, then collapse, as was observed widely in the 1964 Niigata earthquake in Japan. An important difference between the situations in 1934 and today is that the population throughout the Ganges valley is now not only much bigger, but is concentrated in large cities and living in poorly constructed multi-storey apartment blocks. Of particular concern is the obvious 'gap' in earthquakes along the Himalayan front north of Delhi, where no major event has occurred since at least 1500 and where the concentration of major cities is particularly dense. There is little doubt that earthquakes which once killed a few thousand would now kill many more, perhaps hundreds of thousands or more, when they recur in the future, as is inevitable. Thus the earthquake at Muzaffrabad in 2005, though by no means the biggest known earthquake along the Himalayan front and certainly not in the most densely populated part of it, killed many more people than any of the previous known earthquakes in Figure 6.8. The lesson of Muzaffrabad is that there is worse to come.

Tsunamis and the earthquake cycle

The Sumatra–Andaman earthquake of 2004 was the second largest on Earth in the last 100 years; only the 1960 Chile earthquake was larger. To put this massive event into some perspective, the fault that moved was 1200 km long (a distance from London to Rome) and slipped up to 25 metres. By comparison, the fault that destroyed Bam in Iran in 2003 was 12 km long, and moved about 1 metre. The time taken to tear the fault from one end to the other took 7 minutes in Sumatra, compared with 7 seconds in Bam. As in all the previous examples, the fault in the Sumatra earthquake pushed one block (Sumatra) on top of another (the Indian Ocean), and is the same type of fault that occurs round most of the edge of the Pacific as well.

Most of the people who perished in this earthquake died in the tsunami, and most of those lived close to the fault that moved, in Sumatra, though several thousand were killed when the tsunami hit the more distant shores of Thailand, Sri Lanka, the Maldives and East Africa. To appreciate what happened, and the consequences for the future, one must first understand the nature of the earthquake cycle, illustrated in Figure 6.9. The Indian

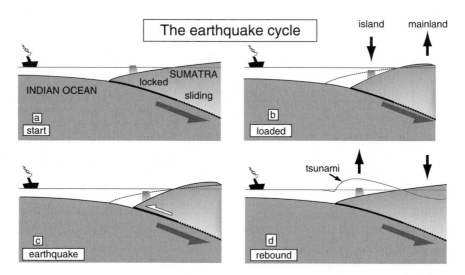

FIGURE 6.9 The earthquake cycle. (a) As the Indian Ocean slides eastwards beneath Sumatra at about 60 mm per year (big arrow), the fault (solid black line) separating them is locked by friction above 40 km depth. Below this depth (dashed line) the rocks are sufficiently hot and weak to slide continuously. (b) As the locked fault is dragged down by the steady movement of the Indian Ocean, the island adjacent to fault is submerged, while the mainland coast is bulged up. Their original levels in (a) are shown by the dashed line. At this stage the fault is 'loaded' and ready to slip in an earthquake. (c) When the earthquake happens, the fault is released (white arrow). (d) The sea bed returns to its original level in (a), but this happens so quickly that the sea surface itself is offset, creating a broad wave, or tsunami, which then flows away in both directions. The cycle then repeats, with about 200–300 years between earthquakes.

Ocean slides beneath Sumatra at a steady rate of about 60 mm per year (Figure 6.9a). The fault is the interface between them, but is locked by friction and does not move at depths less than about 40 km; below that depth the rocks are sufficiently hot and weak and can slide past each other steadily. Since the fault is locked, it is dragged down by the moving Indian plate, causing the island near the fault to be submerged, while the adjacent mainland is bulged up and raised higher (Figure 6.9b). Eventually the strain accumulated on the fault overcomes the frictional resistance and the fault is released (Figure 6.9c), uplifting the island and sinking the mainland, so that they return to their original positions (Figure 6.9d), and the whole cycle can then repeat again (back to Figure 6.9a). In Sumatra the repeat time between earthquakes is around 200–300 years. When the fault moves in the earthquake, the land surface above the fault is restored very quickly, changing elevation by up to 5 metres in about 5 seconds. As a result, the sea surface also changes quickly, uplifted near the island and dropped near the mainland, forming the wave known as a 'tsunami'. The wavelength of the tsunami (i.e. twice the distance from peak to trough) is about 300 km, much greater than the average water depth of the deep oceans (about 5 km), and it is this property that allows the tsunami to travel round the world with little change in shape and a speed of around 800 km per hour.

This basic process, and the properties of such tsunamis, have been well understood for some time. Since the wave keeps its shape as it travels, people to the west of the fault (left side of Figure 6.9), in Sri Lanka, the Maldives and East Africa, experienced a big up-pulse followed by a down-pulse. Those in the east (right of Figure 6.9), in Thailand and coastal Sumatra, experienced first a down-pulse, or withdrawal of the sea, followed by an up, or large incoming wave. Even travelling at 800 km per hour, it takes 15–20 minutes for the down pulse of the wave to travel its length before the high incoming wave that follows it can arrive. After the event, many newspapers showed shocking pictures of tourists in Thailand who went down to the beach to examine the exposed sea bed, with its mud, old wrecks and stranded fish; some of these people were drowned when the incoming high wave followed 15–20 minutes later.

From this simple account, it is clear that what matters in this situation is education, or knowing what to do. If the sea withdraws, you have about

10–15 minutes to run inland and try to reach higher ground. If you are on the beach and feel a large earthquake, it is best to assume there will be a tsunami to follow, and run inland without delay. Around the Pacific, in south, central and north America, in Japan, the Philippines, and New Zealand this lesson is repeated frequently to schoolchildren and, because such earthquakes and tsunamis are relatively frequent, the general level of awareness is high. In the Indian Ocean, such events are much rarer, and most people did not know what to do. There were, however, exceptions. The people of Simeulue Island, west of Sumatra and in the position of the island in Figure 6.9, did know what to do, because of an earlier earthquake in 1907, the memory of which had been passed down through older people. On that island only seven people were killed out of 75 000, compared to well over 200 000 killed in northern Sumatra. Similarly, in some of the Andaman and Nicobar islands north of Sumatra, the indigenous ancient tribes knew what to do, and survived, while more recent immigrants from the Indian mainland did not, and drowned.

As in the earlier examples from Iran, the earthquakes affect and control the landscape on a timescale that is relevant to human occupation. Figures 6.10a and 6.10b show part of Simeulue Island, whose coast was uplifted in the rebound after the 2004 earthquake (Figure 6.9d): the chequerboard pattern exposed in the mud flats now uplifted above high-tide level is formed by old rice paddy fields that were submerged and abandoned in the sinking, or loading, phase of the earthquake cycle (Figure 6.9b), probably several generations ago. Now they have been returned to an elevation that allows them to be cultivated again. The December 2004 earthquake was followed by another, also very big earthquake, in March 2005, further south-east along the coast of Sumatra. Figure 6.10c shows part of the island of Nias, taken in January 2005, when the island had been submerged by the fault-loading phase (Figure 6.9b), drowning an earlier coconut plantation. After the March 2005 earthquake, the land rebounded (Figure 6.10d) returning the once-drowned plantation to a position above high-tide level. These examples show that it is not surprising that a rudimentary knowledge of the effects of large earthquakes should survive in coastal populations, providing they are stable and sedentary enough for such experience to accumulate as folklore. Observations similar to those in Figure 6.10 show us that the fault further south-east of the 2004 and

FIGURE 6.10   Observations of the earthquake cycle on human timescales.
(a) Aerial photo of Simeulue Island, taken after the December 2004 earthquake. The island is in the position shown in Figure 6.9b, and has rebounded
up after the earthquake (Figure 6.9d) lifting it out of the water. High-tide
level before the earthquake was at the base of the trees; the mud flats are
now exposed at all stages of the tide. (b) Close-up of part of the mud flats near
(a): the chequerboard pattern reveals rice paddy fields, originally submerged
in the loading phase of the cycle (Figure 6.9b) and abandoned, but now re-
exposed after the earthquake. (c) Part of the coastline of Nias Island, before
the March 2005 earthquake. High-tide level is at the beach near the palm
trees. Note the regular pattern of tree stumps in the water in the centre of
the picture: these are the remains of a coconut plantation, dragged beneath
the sea and killed during the loading cycle (Figure 6.9b). (d) View of the
same area as (c) taken after the March 2005 earthquake: the entire region
has rebounded and been uplifted by the earthquake (Figure 6.9d), so that
the coconut plantation is now exposed at all stages of the tide. This picture
was taken near to high tide. (All photos courtesy of Kerry Sieh.)

2005 earthquakes, closer to the urban centres of Padung and Djakarta, is
loaded and ready to break sometime in the future; this story from Sumatra
is not over yet.

In the aftermath of the 2004 disaster, there was much talk of tsunami
warning systems in the ocean. These exist already in the Pacific, in the

form of ocean-bottom pressure sensors that can detect a tsunami wave as it passes over them, and issue an alert via satellite-based communications. Such a system can give warnings of a few hours to places a few thousand kilometres distant from the earthquake source area, such as Hawaii. They were installed to reduce false alarms; not all oceanic earthquakes generate tsunamis that will travel large distances, and seismological information alone is rarely sufficient, or can be processed quickly enough, to be a reliable discriminant between them. Evacuation of coastal regions is a difficult and expensive business, and false alarms reduce the willingness of coastal populations to heed warnings. In the case of the Sumatra–Andaman earthquake, such systems would probably not have been able to give sufficient warning in Thailand (which the tsunami reached after 40 minutes), might have been helpful in Sri Lanka (2 hours), probably would have helped in the Maldives (3.5 hours) and certainly would have helped in East Africa (8 hours), if accompanied by effective contingency plans. But such warning systems would have been no use at all for the local coastal populations in Sumatra, who sustained by far the majority of the casualties, and who had less than 20 minutes after the earthquake before the tsunami hit. A necessary condition for saving lives in that situation is knowing what to do. That, of course, may not be sufficient in itself: in some low-lying coastal and delta regions there may be no high ground nearby, and some attention to engineered infrastructure may be necessary, in the form of constructing evacuation routes, high stable structures that will allow a wave to pass beneath them ('vertical evacuation'), or even barriers to deflect or reduce the initial impact of the wave: these are all aspects that are being addressed in the circum-Pacific, and much expertise is available. Nonetheless, the effectiveness of education in tsunami-prone regions is so clear that it should be given very high priority.

Earthquake vulnerability in the modern world

The developing world has seen a relentless rise in population and its concentration into large towns, cities and megacities. In the great earthquake belts of Asia, many of these concentrations are adjacent to mountain fronts and faults, largely because of their agricultural origins. Yet their development has not been accompanied by a decrease in their vulnerability to earthquakes. The existence of building codes in many of them, poorly enforced or observed, has had little effect in many of the countries concerned, and mortality rates remain shockingly high: for example, 20–35% of the population (between 240 000 and 500 000) died in the 1976 Tangshan earthquake in China. Half the world's mega-cities of more than 10 million inhabitants are in locations vulnerable to earthquakes, and events that in the past killed a few hundred or thousand people will now kill tens or hundreds of thousands, or more. The reason we have not yet had an extreme catastrophe, with over a million killed in one earthquake, is only because these cities have been exposed for a short time (about fifty years) compared with typical repeat times of earthquakes on faults (usually hundreds or thousands of years). But a catastrophe of those dimensions seems to me to be inevitable, and will probably occur in this century.

Meanwhile, the developed nations have had great success in reducing the earthquake vulnerability of their urban populations, at least for moderate-sized earthquakes. In California, and increasingly in Japan, earthquakes of magnitude 6 to 7, which can routinely kill tens of thousands in rural areas of the Middle East and Asia, are now principally stories about economic loss. The earthquakes at Loma Prieta in northern California (1989; magnitude 7.1, 64 killed) and Northridge in southern California (1994; magnitude 6.8, 50 killed) occurred in regions that would be considered urban or suburban compared with Bam (2003; magnitude 6.8, 40 000 killed) or Tabas (1978; magnitude 7.3, 11 000 killed) in Iran. Expressed as proportions of the local populations, the mortality figures of the Californian earthquakes are insignificant (much less than 0.1%) compared to those of the two in Iran (roughly 30% and 85%). This comparison is not to lessen the importance of economic loss, as that can also be devastating for developing countries. The costs of the 1989 and 1994 Californian earthquakes were 0.2% and 1% of the regional (not national) GDP, while the cost of the 1972 Nicaragua and 1986 El Salvador earthquakes were 40% and 30% of those countries' *entire* GDP. The experiences of California and Japan are testaments to good building-design codes, sensibly enforced, though whether those same designs will prove effective in earthquakes much bigger than magnitude 7, in which shaking durations and ground displacements are much larger, is uncertain. Nonetheless, the message that good buildings save lives could not be clearer, and is an issue far more important than demands for earthquake prediction, which has remained scientifically elusive.

The future: what can be done?

What can be done about the appalling earthquake vulnerability of large mega-cities in the developing world? The problem is immense and urgent, often generating a response of despair in local politicians who can see no achievable result with the limited resources at their disposal, which are anyway in demand by other projects. Certainly, it involves preparation for the inevitable extreme catastrophe, so as at least to attempt to cope with its consequences. But it also needs a sustained determined effort at hazard mitigation and reduction *before* such a catastrophe occurs, and one that must embrace a large number of cities. The problem of what to do about populations that are already housed in poorly constructed apartment blocks that accompanied the rapid growth of mega-cities is particularly difficult. One thing at least is clear: we can expect 2 billion people to be added to the cities of developing countries over the next twenty years. This will be the biggest construction boom in history, and for those people at least, we should try and ensure that this construction conforms to good building practice and good land-use planning. In this more limited goal, education plays a key, and deliverable part. In the 1994 Northridge, California earthquake, the modern USC hospital, located near the fault that moved, was completely undamaged and remained functional, with no interruption to operations. The famous quote was that 'Not one test tube was lost.' Patients refereed to 'a gentle rocking' and did not know a major earthquake was occurring until they heard about it later from hospital staff. In the same-sized earthquake of 2003 at Bam, in Iran, all the modern hospitals and clinics ceased to function through collapse or damage. When the public in the developing world start to realise that total destruction is not inevitable, and to demand that their new buildings conform to modern standards in the West, some progress, at least, will have been made.

The next chapter considers another threat to survival whether the risk of mortality is modulated by global inequalities – famine.

FURTHER READING

Ambraseys, N. N. and Melville, C. P. (1982). *A History of Persian Earthquakes*. New York: Cambridge University Press.

Bilham, R. (1998). 'Earthquakes and urban development', *Nature* **336**, 625–6.

Bilham, R. (2004). 'Urban earthquake fatalities: a safer world, or worse to come?', *Seismological Research Letters* **75**, 706–12.

Jackson, J. (2001). 'Living with earthquakes: know your faults', *Journal of Earthquake Engineering* **5**, 5–123.

Smith, A. (1953). *Blind White Fish in Persia*. London: George Allen & Unwin.

Tucker, B. (2004). 'Trends in global urban earthquake risk: a call to the international Earth Science and Earthquake Engineering communities', *Seismological Research Letters* **75**, 695–700.

Wulff, H. E. (1968). 'The qanats of Iran', *Scientific American*, April, 94–105.

Yeats, R. S., Sieh, K. and Allen, C. R. (1996). *The Geology of Earthquakes*. Oxford: Oxford University Press.

7 Surviving famine

ANDREW PRENTICE

Introduction

Famine. Who amongst us has ever contemplated the true horror of that short word? Who amongst us has ever had to endure even a single day without food? Who amongst us has had to worry about where our next meal will come from, or how we will feed our children? Who has had to anguish over whether to sell or abandon their children in the hope that another kind soul will feed them? Who amongst us has ever had to contemplate cannibalism – even of our own family members? Yet the thesis of this chapter is that until very recently famine has been an ever-present threat to the survival of our forebears and a pivotal driver of the natural selection of our species. The first step in developing this argument is to call on historical evidence to provide a measure of the frequency and intensity of past famines, and of their associated mortality. This will be followed by a description of the biological and behavioural adaptations that are brought into play as famine progresses in an attempt to ensure survival until salvation arrives. Finally I will consider the contemporary implications of our forebears' struggle against hunger in terms of modern disease patterns.

The historical record of famines

Ancel Keys, author of a two-volume treatise on *The Biology of Human Starvation*, wrote that 'The causation of famine by unkind nature was by

Survival, edited by Emily Shuckburgh. Published by Cambridge University Press.
© Darwin College 2008.

FIGURE 7.1 Famine. (Photo courtesy of Tom Stoddard, Getty Images.)

far the most common until fairly modern times, when man's dominance of nature allowed him to assume the role of creator of his own misery.' This division between natural and human-generated famines will be adopted below as we traverse continents and millennia in a brief history of famine. There is an extensive literature on the definition of famine, but for current purposes it can be taken to mean a widespread shortage of food among large parts of a population that is associated with significant mortality from starvation (see Figure 7.1).

Famines caused by unkind nature

For those who follow the Christian scriptures the Book of Genesis represents the beginning. Genesis contains twenty-four references to famine, such as this: 'And there was no bread in all the land; for the famine was very sore, so that the land of Egypt and all the land of Canaan fainted by reason of the famine.' Although these references are to a single calamitous event, the frequency of repetition and the centrality of famine to the events described underlines its importance in the consciousness of the writer.

The flood plain of the Nile in Upper Egypt has always been vulnerable to famine because its fertile soils and high crop yields supported a high population density, becoming the very cradle of civilisation when times were good. But all of this depended on the annual floods. One of humanity's oldest known written records is the Stele of Famine (approximately 2000 BC) in which the emperor Tcheser records:

> I am mourning on my high throne for the vast misfortune, because the Nile flood in my time has not come for seven years. Light is the grain; there is a lack of crops and of all kinds of food. Each man has become a thief to his neighbour. They desire to hasten and cannot walk... The counsel of the great ones in the court is but emptiness. Torn open are the chests of provisions, but instead of contents there is air. Everything is exhausted.

And Egypt provides many other records of famine both written and pictorial (see Figure 7.2). A Middle Kingdom hymn to the Nile laments that 'If there be a cutting down of the food offerings of the gods, then a million men perish among mortals, covetousness is practised, the entire land is in a fury, and great and small are on the execution block' and the Sepulchers of Ankhtifi describing the famines of 2180–2160 BC tell how 'All of Upper Egypt was dying of hunger, to such a degree that everyone had come to eating his children... the entire country had become like a starved grasshopper...'

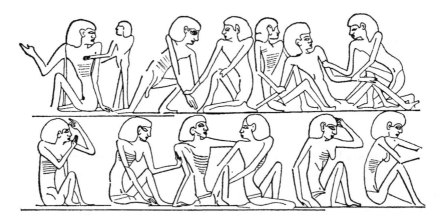

FIGURE 7.2 Famine victims from the first half of the third millennium BC depicted on the causeway at the Pyramid of Unas at Sakkara.

In *Floods, Famines and Emperors* Brian Fagan argues that changes in weather patterns ranging from short-term El Niño shifts to the Little Ice Ages caused the downfall of many great civilisations, from the Pharaohs of Egypt to the Moche of Peru, and from the Mayans of Tikal to the Anasazi of North America. These civilisations arose in times of plenty when the climate favoured rapid population growth. But it only took a few years of drought, or a single massive storm to sweep away the topsoil, and starvation led to the irreparable collapse in population numbers.

The Roman Empire was afflicted by famines, with records of thousands of starving citizens committing suicide by throwing themselves into the Tiber in 436 BC. And in the late Roman and early Byzantine empire an exhaustive study of the historical evidence has identified 121 famines and 101 epidemics between the fourth and the seventh centuries.

Moving to the Indian subcontinent we witness some of the world's greatest ever hardships due to the region's dependence on the monsoon rains. Describing Kashmir in AD 917–18 Kalhana's *Rajatarangini* records that 'One could scarcely see the water in the Vistasa, entirely covered as the river was with corpses soaked and swollen by the water in which they had long been lying. The land became densely covered with bones in all directions until it was like one great burial ground, causing terror to all beings.' An inscription in the Narasimhaswamy Temple at Kadiri referring to the famine of 1390–91 records 'Innumerable skulls were rolling about, and paddy could not be purchased even at the cost of 10 Nali per one Panam.' And inscriptions describing a fifteenth-century Deccan famine report that 'for two years no grain could be seen and in the third when the Almighty showered his mercy upon the earth, scarcely any farmers were left to cultivate the lands'.

China too has been the home to great famines. Mallory's classic *China: Land of Famine* estimated that in one region or another there were 1828 famines between 108 BC and AD 1911 – nearly one every year.

But these are famines in far-away lands. Like today's famines in Darfur or Niger they may seem to have little relevance to European and American readers. Have our own forebears really been savaged by famines in a similar way? The answer is a categorical 'yes'. In an appendix to *The Biology of Human Starvation*, Ancel Keys and colleagues list historical evidence for 175 famines in the British Isles since Caesar's invasion. Many of these have references to very high mortalities and to cannibalism which validates their severity. The 'micla hunger' of AD 976, the decade of famine following the Norman conquest in 1066, and the Great Famine of 1315–21 are some of the best-known examples, but there are scores of others.

In *The Great Famine: Northern Europe in the Early Fourteenth Century*, W. C. Jordan describes the devastating effects of starvation across a wide area of north-west Europe including the British Isles: 'A calamity unheard of among living men.' It is estimated that between one-third and one-half of the population perished. According to a contemporary report from Flanders at the beginning of the famine in 1319, 'Parents killed their children and children killed parents, and the bodies of executed criminals were eagerly snatched from the gallows.'

And then there was the best-documented famine the world has known – the Irish potato famine of 1845–49. Among the scores of excellent books written about this particular famine Cecil Woodham-Smith's *The Great Hunger: Ireland 1845–49* probably gives the fullest description of the administrative and political dimensions to the catastrophe, and contains some of the most heartrending descriptions. A letter from Capt Wynne to his commanding officer in County Mayo on Christmas Eve 1846 contains the following passage:

> I ventured through the parish this day to ascertain the condition of the inhabitants, and, altho' a man not easily moved, I confess myself unmanned by the intensity and extent of the suffering I witnessed especially among the women and the little children, crowds of whom where to be seen scattered over the turnip fields like a flock of famished crows, devouring raw turnips, mothers half naked, shivering in the snow and sleet, uttering exclamations of despair while their children were screaming with hunger. I am a match for anything else I may meet here, but this I cannot stand.

The Irish Famine was certainly grievous and, for those who tried to flee, its effects were compounded by the horrors of the 'coffin ships' for which unscrupulous ticket-brokers sold passages on unseaworthy vessels and cheated the vulnerable of their rations and water supplies causing massive further mortality before the ships docked in America.

Elsewhere I have considered the possible effects of these enforced migrations from famine on the endowment of the genetic stock of Irish Americans

(Prentice 2001), and before leaving this topic it is worth recalling the fate of the Pilgrim Fathers – not because they have a significant effect on America's genetic make-up, but because their story provides an annual reminder to Americans of their forebears' struggles against famine. From the 104 passengers who set sail on the *Mayflower* on 16 September 1620 only 23 left any descendants. They suffered terrible mortality during the first two winters when their inexperience of the new conditions led to almost total crop failures and the loss of all their seed corn. In his journal Governor Bradford wrote that 'The fales of their grounds which came first over in the Mayflower according to their loses were cast...The greatest halfe dyed in the general mortality.' And today the feast of Thanksgiving commemorates the deliverance of those few survivors.

Famines caused by man's inhumanity to man

Twenty-five years ago in his book *Poverty and Famines: An Essay on Entitlement and Deprivation*, Amartya Sen challenged the view that famines are usually simple food crises caused by drought, pests and crop failures. He demonstrated that they are often, perhaps usually, economic disasters, and developed his 'food entitlement theory' which better explains why it is the poor that die in famines. He reminds us that 'Famines imply starvation, but starvation does not necessarily imply famine.' Centuries previously the famines in England between 1066 and 1638 had been attributed not only to the weather, but also to the 'abasing of the coin' and the 'uncharitable greediness and unconscionable hoarding of corn master and farmer'.

A few examples of how man has become 'the creator of his own misery' will suffice to reinforce the argument that famines have been an ever-present threat. The Second Book of Kings in the Bible records Benhadad's siege of Samaria when 'an ass's head was sold for fourscore pieces of silver, and the fourth part of a cab of dove's dung [*colloq.* 'a small pot of grain'] for five pieces of silver'. We shall tragically return to this story later.

Some of the greatest man-made famines have occurred in the twentieth century due to the greater size of populations and the greater efficiency of instruments of subjugation. One of the worst was Stalin's Terror Famine of 1932–33 in the Ukraine and neighbouring Cossack lands. All food stocks were forcibly requisitioned, a military cordon (the 'ring of steel') prevented food supplies from entering, and the population was left to die in a famine organised as an act of state genocide. In *The Harvest of Sorrow*, Robert Conquest estimates that 7 million Ukrainians died – '20 souls for every letter of this book'.

But probably the greatest famine the world has known was caused by Mao's 'Great Leap Forward'. China had suffered previous famines in the twentieth century. It had major famines in 1916 and in 1929 when 2 million people died in Hunan Province alone. The Japanese invasion and the subsequent Kuomintang rule heralded further starvation with grain prices increasing seventy-fold overnight and inflation rising to almost 3 million percent by the end of 1948. But none of this compared with the catastrophic mortality caused by Mao's policies when peasants were forced to melt their agricultural implements and cooking pots to make steel, and when collectivisation and misguided state advice about how to increase crop yields caused massive shortages leading to an estimated 30 million deaths from starvation. This was at a time when Mao was exporting grain to Russia to repay a previous loan. Readers of Jung Chang's *Wild Swans* may recall a chapter entitled 'Capable women can make a meal without food' – a grotesque Maoist corruption of the ancient Chinese saying 'Even a capable woman cannot make a meal without food.' Jung Chang tells of her parents eating chlorella grown in their own urine as a cure for famine oedema (the build-up of excess fluid in body tissues leading to a puffy appearance), and of the executions of traders caught selling the meat of babies and children as wind-dried rabbit.

Seasonal famines

So far I have described episodic famines of catastrophic proportions that hit with little warning. There is another type of hunger, often less severe, but which, by virtue of its regular repetition, has probably had an equal impact on human selection. These are the annual 'hungry seasons' still suffered by many populations to this day. Most of my current research is conducted within one such population in rural Gambia where a single rainy season dominates the calendar of life. During these rains all able-bodied members of the community work long hours at subsistence farming.

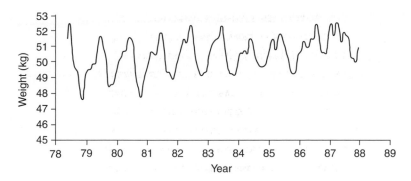

FIGURE 7.3 Annual swings in body weight in rural Gambian women caused by hungry and harvest seasons.

By definition this is the time of year furthest from the previous harvest and prior to the next harvest. In most years food supplies start to run out and the combination of high levels of energy expenditure and low levels of intake cause a seasonal weight loss that usually amounts to about half of peoples' body fat stores (see Figure 7.3). A local Mandinka proverb goes *'Nin i kono fata, i si kumbo, kaatu fo i si na konko'* ('When your stomach is full, you should cry, as you will shortly go hungry'). This is fat acting as nature intended, to provide a buffer against times when food is short, and we shall examine below how this might influence genetic selection through a profound seasonal variation in fertility rates.

Although modern societies have escaped from seasonal hunger through advanced means of food preservation, storage and transport, such was not always the case. Lacey and Danziger's account of *The Year 1000: What Life Was Like at the Turn of the First Millennium* records an exactly analogous situation in Britain in which summer food shortages were relieved by the first harvest celebrated on 1 August as Lammas Day (*loaf-mass day*) (see Figure 7.4). William Langland's *Piers Ploughman* gives us a flavour of what it was like in the fourteenth century with quotes such as 'Then hunger seized him by the belly, wringing his guts until the water ran from his eyes...' and with clear descriptions of the struggle to survive each hungry season. At one point Piers states wistfully 'that with these few things we must live till Lammas time, when I hope to reap a harvest in my fields...Then I can spread you a feast as I'd really like to.'

FIGURE 7.4 Representations of seasonal hunger in medieval Britain. (Reproduced with permission from Lacey R. and Danziger D. (2000). *The Year 1000: What Life Was Like at the Turn of the First Millennium*. London: Abacus.) © 1999 by Robert Lacey and Danny Danzinger. By permission of Little, Brown and Co., Inc.

The evolutionary timescale of famines

The general thinking is that hunter–gatherer peoples would have had short periods of hunger but would have been unlikely to suffer starvation except when incapacitated. Hunter–gatherers lived in small groups, had a very diverse diet and could move rapidly when local resources became depleted. Famine, it is argued, is a recent phenomenon paradoxically created by the very discovery that has allowed population growth and the formation of civilisations – namely agriculture. Around 12 000 years ago humans started to learn how to grow corn and other grains and gradually become more and more dependent on these staples with an accompanying restriction in dietary diversity. When climate is stable this strategy succeeds, but as Brian Fagan and Jared Diamond and others have shown this dependency creates great vulnerability to famine.

A pertinent question in light of the above is whether 12 000 years (or about 600 generations) is sufficient time for evolutionary selection to have occurred. Having spent many years studying the energetic adaptations that Gambian women can employ in order to sustain reproduction against all the odds of a poor diet and seasonal hunger, and having studied women who at menopause have ten live offspring and others who have none, I am a firm believer that 600 generations provides ample scope for the selection of important energy-sparing survival traits, and my research group has described the natural biology of some of these adaptations in some detail.

Famine's imprint on the human genome

In *On The Origin of Species* Charles Darwin noted: 'A grain in the balance shall determine which individual shall live and which shall die.' His reference to a grain was no doubt to the unit of weight, but in the context of famine the phrase develops a fresh poignancy if we interpret it as a grain of wheat or rice. In the final paragraph of that book Darwin also stressed the likely impact of famine as an instrument of natural selection: 'Thus, from the war of nature, from famine and death, the most exalted object which we are capable of conceiving, namely, the production of the higher animals, directly follows.'

In February 2001 the journals *Nature* and *Science* simultaneously published the first draft maps of the human genome – the 800 Bibles-worth of information contained within each nucleus of our body. This blueprint has been moulded by aeons of adjustments to optimise survival and our ability to pass on this endowment to future generations. Can we detect signals of selection by famine within this map? Can we identify genes that equip us with metabolic or behavioural traits that enhance our survival when the granary is bare?

The search for such genes started forty years before the publication of the human genome with the work of J. V. Neel at the University of Michigan. Neel was fascinated by various aspects of the metabolism of diabetic women, especially the abnormally fast growth of their fetuses. In 1960 he published a paper entitled 'Diabetes mellitus: a "thrifty" gene rendered detrimental by "progress" ', and since then his concept of thrifty genes has been a touchstone for biologists trying to understand the modern epidemic of obesity and diabetes. In the early days Neel's disciples frequently invoked the theory to explain the extraordinarily high rates of diabetes among many Polynesian islanders. They argued that the islands were populated by a few founder members who were the rare survivors of long sea voyages during which many died of hunger. Goldie's painting of the *Arrival of the Maoris in New Zealand* (Figure 7.5) seemed to me the perfect representation of just such an event. However, Maoris themselves view the painting as ethnographically inaccurate and an insult to their ancestors. They proudly claim that their ancestors had a complete mastery of the oceans and could sail as far as they wished surviving on fish and

FIGURE 7.5 *Arrival of the Maoris in New Zealand* by R. A. Goldie. (Reproduced with permission from Auckland Art Gallery Toi o Tamaki, gift of the late George and Helen Boyd, 1899.)

rainwater. Far from ruining a good story, this claim formed the impetus for the deeper research into the history of famines that forms the basis of my current thesis that we *all* carry the imprint of our ancestors' survival in the face of famines.

The concept of thrifty genes remains a nebulous one, but is probably no less useful for being so. Indeed it is the very imprecision of the concept that has ensured its popularity and survival. Nobody has yet shown convincing evidence for the existence of a thrifty gene, yet the basic idea that we have an ancient metabolism tuned to saving energy when supplies are short seems self-evidently plausible. So long as it is not interpreted too literally it will continue to serve as a useful working hypothesis. Moreover, the next chapter will explore evidence of genetic control of this type across the animal kingdom.

Mechanisms for genetic selection

When considering the survival of the fittest most of us probably assume that the process is largely driven by the greater mortality of weaker individuals before they have completed their reproductive lifespan. Such effects are indeed important, but in the case of famine they might not represent the most powerful selection effect. This is because starvation has a major effect on reproductive function in both men and women, and because even in the non-starved state there is a greater attrition between when the ovum is fertilised and the birth of a baby than there is postnatally even under conditions when infant, childhood and young adult mortality rates are very high. Thus we should also examine factors that can influence differential rates of conception and fetal survival if we are to fully understand the selective effects of famine.

In the Dutch winter famine of 1944–45 midwives saw their clientele dwindle from 250 deliveries per week to 85. In the Siege of Leningrad the birth rate in the State Pediatric Institute dropped from 447 per month in 1940 to 13 per month in 1942. In the great famine in Madras in 1877 there were only 39 births in relief camps caring for 100 000 people over many months and in the surrounding regions the birth rate fell from 29 per 1000 members of the population down to just 4 per 1000. During the Chinese famine of 1960–62 calculations show that this suppression of fertility had a greater impact on the population pyramid than actual mortality. The last three examples are extreme cases, but in milder famines parents who could continue to reproduce in the face of hunger would probably pass on more grandchildren than those that could not.

The power of this effect can also be appreciated in contemporary populations that still suffer hungry seasons. Figure 7.6 shows a profound annual swing in the monthly frequency of births in rural Gambia and Bangladesh. The timing of this suggests that it is mediated by a biological suppression of fertility since the nadir occurs exactly nine months after the peak period of maternal weight loss in the hungry season. This is the same phenomenon that causes ovulation to be suppressed in very thin ballerinas, gymnasts and anorexics, and is caused by a decrease in the hormone leptin when fat stores are depleted. The important question in the current context is whether the women who can continue to conceive in the hungry season (those births circled in Figure 7.6) have one or more thrifty reproductive genes which allow them to conceive all year round whilst others can only conceive in the harvest season.

One obvious metabolic trait that might explain such differential sensitivity to food shortage is simply a woman's propensity to lay down fat in

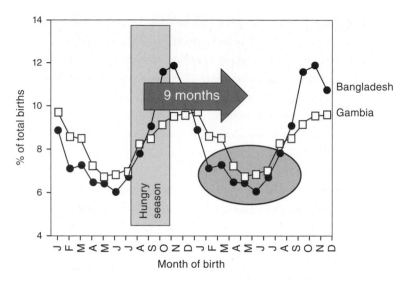

FIGURE 7.6 Seasonal swings in fertility in Gambia and Bangladesh illustrating the suppression of fertility by hunger and weight loss.

the harvest season. Any woman arriving at the onset of the hungry season, or at the onset of a famine, with much greater fat stores than her peers is more likely to remain fertile for longer. This is likely to be the reason that so many ancient fertility dolls such as the Venus of Willendorf (Figure 7.7) have an exaggeratedly 'gynoid' body shape with large fat stores around the thighs, buttocks and breasts. Biochemical studies have shown that these fat depots are the most sensitive to hormonal signals during reproduction and hence are likely to serve a specific purpose in releasing energy for the fetus or to support milk production.

This speculation about a possible relationship between a woman's body fat pattern and fertility leads us to the third way in which famine may have influenced natural selection; namely through the sexual selection by men of women who signal a greater reproductive fitness through their fatness and the anatomical distribution of this fat.

Any of these three methods of selection (i.e. death by famine, differential fertility in times of hunger, and mate selection according to fertility-related cues) could have influenced the famine survival traits discussed in the following sections.

FIGURE 7.7 The Venus of Willendorf. (© Naturhistorisches Museum Wien, Photo: Alice Schumacher.)

Survival strategies

Metabolic survival strategies

Humans stand out from other primates, and from all except the aquatic mammals, as having evolved large fat stores. For instance, wild baboons have about 4% body fat (by weight) whereas European men have about

15–20% body fat and women have about 30–35%. Even the rural Gambians we study have about 12% and 20% body fat in men and women respectively. Fat is the most efficient way of storing large quantities of energy because it provides the best energy-to-weight ratio as well as being a good insulator and providing physical protection against injury. The question as to why humans have made such an unusual evolutionary choice in terms of the amount of fat they store remains unanswered, but the survival value of large energy reserves in the face of seasonal and episodic famine would seem an obvious selective factor.

The deposition and mobilisation of fat in adipose tissue stores is under complex control by a number of hormones among which insulin is a key player. Insulin also controls the uptake of glucose and fat by muscles so the fate of extra energy consumed at times of feasting will be determined by the relative sensitivity of muscle and adipose tissue to the actions of insulin. Neel's original thrifty-gene idea proposed that insulin resistance in muscle (i.e. a low level of glucose uptake for any given amount of insulin in the blood stream) would be advantageous by diverting energy to fat stores at times of excess in readiness for times of hardship. He proposed that it is this inherent insulin resistance in muscle, ultimately leading to diabetes, that is the adaptation 'rendered detrimental by progress' since we are no longer exposed to food shortages.

If building up reserves of fat during the harvest months is the first line of defence then the careful martialling of these resources by inducing energy-sparing adaptations forms the second line. As the body recognises the first stage of starvation it modifies the fuels it consumes in order to use as much fat and as little glucose as possible. This is because in total starvation, when no carbohydrate is provided in the diet, glucose has to be produced from the breakdown of muscle tissue which, if left unhindered, would quickly damage vital organs and lead to an early death.

Energy and protein reserves are also protected by a decrease in the body's rate of burning energy. The most detailed investigation of these metabolic survival mechanisms was undertaken by Ancel Keys and colleagues during a remarkable study of the effects of semi-starvation in thirty-two conscientious objectors towards the end of the Second World War. The purpose of this work was to test the optimal rate of refeeding the tens of thousands of starved concentration camp inmates and prisoners of war that were soon to be liberated by the allied forces. The men were fed half rations for 26 weeks and certainly suffered:

> I'm hungry, I'm always hungry – not like the hunger that comes when you miss lunch, but a continual cry from the body for food. At times I can almost forget about it but there is nothing that can hold my interest for long. The menu never gets monotonous even if it is the same each day or is of poor quality. It is food and all food tastes good. Even dirty crusts of bread in the street look appetizing and I envy the fat pigeons picking at them.

Figure 7.8 shows that at the end of the 26 weeks the men's basal metabolic rate (the essential survival requirements of the body) had declined by 40%. Half of this decrease was caused by the fact that the men had lost 25% of their body weight, and the other half (indicated by the upper curve) was due to a reduction in the energy expenditure of each kilogram of remaining tissue (i.e. to an increase in the efficiency).

Together these adaptations (and others not mentioned here) can eke life out to maximise the chances of rescue by an improvement in the food supply. Sadly we can get a measure of survival times from prisoners who intentionally starve themselves to death. Among the eleven IRA hunger strikers who starved themselves to death in 1982, Bobby Sands survived for 46 days and Michael Devine lasted 73 days. The average was 63 days. It should be remembered, however, that most people enter a famine in an already depleted state (e.g. at the end of a hungry season when the next crop fails) and hence the urgency of aid organisations when it is clear that crops have failed.

These metabolic adaptations have been honed through hundreds of famines to optimise survival for the population as a whole, but for each individual they are a risky business. In saving energy it is necessary for the body to make a short-term gamble that it can cut down on various metabolic 'housekeeping' tasks whilst hanging on for better times around the corner. This is why reproduction is temporarily suppressed. But other compromises carry a high risk – notably a reduction in the amount of energy allocated towards immune defences. Some aspects of immunity are high consumers of energy, notably cell-mediated immunity, and these are

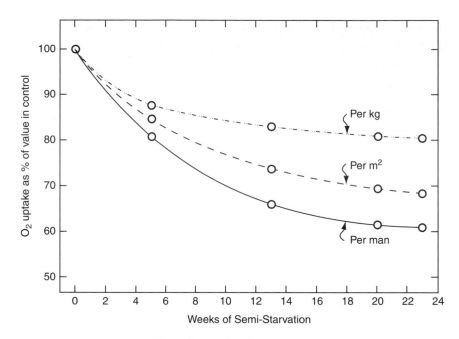

FIGURE 7.8 Effect of 24 weeks of semi-starvation on oxygen consumption of thirty-two men. (Reproduced with permission from Keys *et al.* (1950). *The Biology of Human Starvation*. Minneapolis: University of Minnesota Press.)

compromised in times of starvation; a consequence of which is that deaths from infectious diseases become a major contributor to mortality in times of famine. The pattern of diseases also changes. Tuberculosis becomes a major problem and typhus (frequently known as 'famine fever') can escalate massively. The causative organism for typhus is *Rickettsia* which is transmitted by head and body lice. Thus the combination of lowered immunity and the cramped, unhygienic living conditions of concentration camps, refugee camps, or the Irish coffin ships, create a treacherous combination in which many succumb to typhus before they have finally depleted the last of their energy reserves.

Behavioural survival strategies

Man, being a cognate animal, can add to nature's metabolic survival strategies with a whole spectrum of behavioural adaptations aimed at enhancing

Table 7.1 *Behavioural coping strategies in the face of famine*

Early responses	Desperate measures
Migration of young men to towns	Sale of land –
Changes in cropping patterns	sharecropping – indenture
Dispersed grazing	Break-up of households
Rationing of food	Slavery and serfdom
Inter-household transfer and loans	Theft and petty crime –
Price rises and hoarding	social systems collapse
Sale of possessions	Sale of children (usually
Sale of productive assets	girls) and wives
Consumption of famine foods	Begging and prostitution
	Distress migration – children
	abandoned
	Banditry and civil unrest
	Cannibalism

survival. Some of these may be innate behaviours perhaps driven a subconscious 'memory' of previous famines. Others are carefully planned. They range from benign acts that harm no one to acts of the deepest human depravity induced by a desperation to hang on to the thread of life.

Sociologists and anthropologists have studied these behaviours in some depth and have arranged them in an approximate order according to the severity of the food crisis. Recognition of these patterns of response can be used by aid workers as famine-warning system indicators. Table 7.1 lists some of the choices that face the head of a compound when his family is threatened. They are divided into 'early responses' and 'desperate measures'. Within each of these categories the order with which they are invoked will vary because each household will be faced by a different set of circumstances and will have to respond according to the resources available to them. In ancient China the emperors provided formal guidance to their peasants as to how to order these actions to best survive an impending famine.

Even in poor communities some families may possess easily liquifiable wealth such as women's gold earrings to which they have periodically added small increments of gold when times are good. Many use sheep,

FIGURE 7.9 Famine. (Photo courtesy of Tom Stoddart, Getty Images.)

goats and cattle as their bank account but when everyone wants to sell there may be no buyers and prices will plummet, and in any case when farmers are forced to sell by famine their animals will, by definition, also be in very poor condition. But as desperation sets in people start to sell critical items of living: the doors and windows of their houses and eventually their roofs; their donkeys and oxen and ploughs that constitute their productive assets; and then even their daughters and wives.

This short chapter cannot provide a full discussion of the behavioural strategies but let us examine just a few of the responses in more detail to provide a fuller understanding of the terrors of famine (Figure 7.9).

The consumption of 'famine foods' is a universal feature of the middle stages of crises. Essentially this is a return to the ways of our hunter–gatherer forebears. At first this is simply a rediversification of the diet as people start to consume foods that are perfectly edible but less palatable than their normal fare; domestic and wild animals from dogs to rats are keenly sought. But these are rapidly exhausted and then people begin to consume items that cannot be described as food, but simply as fillers.

Here is a contemporary report from a Swedish famine in 1597, which bears great similarity with others from around the world:

> People ground and chopped many unsuitable things into bread; such as mash, chaff, bark, buds, nettles, leaves, hay, straw, peatmoss, nutshells, peastalks, etc. This made the people so weak and their bodies so swollen that innumerable died. Many widows too were found dead on the ground with red hummock grass in their mouths...

As indicated in the excerpt above the consumption of such foods is highly dangerous because they easily perforate a gut lining that has already been made gossamer thin by starvation. Bacteria enter the bloodstream and cause septicaemia and rapid death.

The decision as to whether to move away from a famine area or to stay put must have tortured thousands of household heads over the course of history. Those of a flexible mind might migrate early on in a crisis. This frequently offers the best chance of survival, but is an enormous decision to abandon one's home and farm and all your possessions. In the Irish famine it was the better off who migrated first; they could afford the fares and perhaps they had relatives in America already. In the first year of the famine thousands went to New York and prospered. But then the authorities panicked at the prospect of the arrival of a poorer type of immigrant in their thousands. Despite the inscription on the Statue of Liberty calling on the world to 'Give me your tired, your poor, your huddled masses yearning to breathe free, the wretched refuse of your teeming shore...' they imposed a prohibitive entry bond which effectively closed their borders to all but the very wealthy. The famine ships turned north to British North America, to St Lawrence and Quebec. There occurred one of the most shameful tragedies in the history of North America as 45 000 starving and fevered émigrés from Ireland were quarantined in a line of ships 2 miles long without food or water until finally the authorities caved in. The eventual capitulation started a continuous procession of ships to Grosse Isle where 'Hundreds were literally flung on the beach, left amid mud and stones to crawl on the dry land as they could' and where boatloads of dead were taken four times a day from single vessels. This one example highlights the difficulty surrounding the decision of whether to flee or to sit tight and hope for relief in one form or another. Soon the door is shut on this decision as the lethargy and physical weakness of starvation removes any option of migration. We can speculate as to whether famine might have bred into the human genome a propensity for flexible thinking and a willingness to migrate to fresh lands and foreign cultures.

As famine deepens the normal rules of society start to crumble. Theft becomes commonplace and is committed without pity or conscience. Tom Stoddart's iconic picture (Figure 7.10) captures one such incident. The journalist John Sweeney wrote of this picture: 'A barefoot African boy crawls in the dust, his whole body, as thin as sticks, keening with misery at the sight of his bag of maize disappearing. He had queued for hours and is on the verge of death, but now his maize has been robbed from him by a fit man who strides confidently away.' Tom Stoddart described how 'The boy's head is tilted, as if asking a question: now that I am dying of hunger, how could someone steal my food?' Yet surveys carried out after famines have shown that almost everyone will admit that hunger drove them to theft.

FIGURE 7.10   Famine. (Photo courtesy of Tom Stoddart, Getty Images.)

And as famine deepens further families are torn apart. Children are given away or sold in the hope that they will survive better with others. This is not generally an act of irresponsible parenthood, far from it. It is generally an act combining elements of sacrifice and sheer desperation. Babies are left in cardboard boxes in hospitals like one in Lanzhou with a note on a scrap of old newspaper reading: 'To kind-hearted people, please look after her. From a mother who regrets her faults.' (Becker 1996.) Other infants are left by the roadside. In the yellow-earth country of north-west China one famine survivor told the following story:

> Those who still had the strength left the village begging and many died on the road. The road from the village to the neighbouring province was strewn with bodies, and piercing wails came from holes on both sides of the road. Following the cries, you could see the tops of the heads of children who were abandoned in those holes. A lot of parents thought their children had a better chance of surviving if they were adopted by somebody else. The holes were just deep enough so that the children could not get out to follow them but could be seen by passers-by who might adopt them.
>
> (Kane 1988.)

How many of these children were adopted, and hence whether or not this was a successful survival strategy, we shall never know.

In the *Heart of Darkness* Joseph Conrad wrote:

> No fear can stand up to hunger, no patience wear it out, disgust simply does not exist where hunger is; and as to superstition, beliefs, and what you may call principles, they are less than a chaff in the breeze. It's really easier to face bereavement, dishonour, and the perdition of one's soul than this kind of prolonged hunger.

And so it is that famine causes many of its victims to break one of humanity deepest taboos – cannibalism. Although it may be practised by only a tiny fraction of people at the depths of famine, stories and evidence of cannibalism occur in all cultures and from the distant past to the present day.

In *Hungry Ghosts: China's Secret Famine* Jasper Becker has drawn together from personal testimonies and official records some of the evidence of cannibalism in the famine of 1960–61. Many of the individual episodes are difficult to authenticate, but the totality of the evidence from this and other famines is overwhelming. And most shocking of all are the abundant reports of parents consuming their own children. We can return to the story of the siege of Samaria in the Second Book of Kings for one of the most heartrending accounts sparingly told in the language of the Old Testament:

> And as the king of Israel was passing by upon the wall, there cried a woman unto him, saying, Help, my lord, O king. And the king said unto her, What aileth thee? And she answered, This woman said unto me, Give thy son, that we may eat him today, and we will eat my son tomorrow. So we boiled my son, and did eat him: and I said unto her on the next day, Give thy son, that we may eat him: and she hath hid her son.

And in China 2000 years ago as the Han Dynasty was founded it is recorded that nearly half the people in the empire died of starvation. This prompted the founding emperor Gao Zu to issue an official edict in 205 BC authorising people to sell or eat their children if necessary.

Over 2 millennia later his words were still being obeyed in China where peasants practised the tradition of swapping their children with their neighbours to alleviate their hunger and to avoid eating their own off-spring. Such behaviour seems absolutely unimaginable to us, but in truth it is arguably the natural biological response. If pregnant mice are exper-imentally starved they will resorb their litters in early gestation, or abort and eat the fetuses in late gestation. If the pups have already been born when the dams run out of food they will eat their pups, then bide their time and start another litter when their food resources improve. If we accept the Dawkinsian argument that even humans are simply here to pass on as many copies of our DNA as possible then, unpalatable as it is to the humanistic or religious mind, consuming one's offspring is a logical strategy for any woman except those close to or beyond menopause.

Individual survival traits

How would each one of us fare in a famine? Who amongst us has the qualities of survival? Could we endure the agony and devise a survival strategy for ourselves and our loved ones? In the comfort of living rooms it might be easy to imagine that we could, but G. B. Leyton, a medical officer captured and interned in a prisoner-of-war camp in the Second World War wrote that 'None of the other hardships suffered by fighting men observed by me brought about such a rapid or complete degeneration of character as chronic starvation.'

To examine the personal qualities that have aided survival we can draw on the words of Primo Levi, one of the very few long-term survivors of Auschwitz who recorded his experiences so movingly in *If This Is a Man* (published in the United States as *Survival in Auschwitz*). Levi observed that 'There comes to light the existence of two particularly well differentiated categories among men – the saved and the drowned' and he offers wonderful character vignettes to explain the qualities that differentiate the two. What follows is a précis of his own words:

In Auschwitz, in 1944, of the first Jewish prisoners only a few had survived – not one was an ordinary Häftling (prisoner). There remained only the doctors, tailors, shoemakers, musicians, cooks, young attractive homosexuals, friends or compatriots of some authority in the camp; or they were particularly pitiless, vigorous and inhuman individuals installed in the post of Kapos, Blockältester, etc.; or finally those who, without fulfilling particular functions, had always succeeded through their astuteness and energy in successfully organising themselves. To sink is the easiest of matters: it is enough to carry out the orders one receives, to eat only the ration, to observe the discipline of the work and the camp. Experience showed that only exceptionally could one survive more than three months in this way. All the musselmans (the Jews' nickname for the 'drowned' souls) have the same story – or rather no story; they followed the slope to the bottom, like streams to the sea. These are the drowned, their life is short but their number is endless. They march and labour in silence, the divine spark dead within them, already too empty to really suffer. One hesitates to call them living, one hesitates to call their death death, in the face of which they have no fear, as they are too tired to understand.

If the drowned have no story, and single and broad is the road to perdition, the paths to salvation are many, difficult and improbable: Primo Levi himself was a chemist who worked in the warm laboratory thus avoiding hard labour and receiving extra foods.

The story of Alfred L shows how vain is the myth of original equality among men. In his own country he was the director of an extremely important chemical factory. He survived by understanding that it is a short step from being judged powerful to being effectively so. His plan was a long-term one in which he took every care not to be confused with the mass. His egotism was absolute. He survived through the principle of 'to him who has it shall be given'.

Elias Linden was a dwarf but he had a bestial vigour – nothing seemed impossible to him. He knew how to make a spoon from a piece of tin, and a knife from a scrap of steel; he found dry paper, wood and coal everywhere and knew how to start a fire in a few moments even in the rain. He could ingest ten, fifteen, twenty pints of soup without vomiting and without having diarrhoea, and begin work again immediately afterwards. Elias was naturally and innocently a thief: in this he showed the instinctive astuteness of wild animals.

Henri, on the other hand, was eminently civilised and sane and possessed a complete and organic theory of how to survive in Lager (the camp): organisation, pity (eliciting pity) and theft. After his brother died, Henri cut off every tie of affection; he closed himself up, as if in armour, and fought to live without distraction with all the resources that he could derive from his quick intellect. He was constantly intent on his hunt and his struggle, hard and distant, the enemy of all, inhumanly cunning and incomprehensible like the Serpent in Genesis.

And Levi's mention of Genesis brings us to another example of an inherent trait that favours the survival of an individual and those associated with them in famine – namely beauty. Except in the most all-embracing of famines there will generally be someone who will care for the beautiful as we are reminded in the story of Abram's journey into Egypt: 'And there was famine in the land: And Abram went down into Egypt to sojourn there: For the famine was grievous in all the land.' Abram pretended that Sarai, his wife, was his sister as she was very beautiful and he feared the Egyptians would kill him and claim her if they knew she was his wife. The story continues: 'And it came to pass, that, when Abram was come into Egypt, the Egyptians beheld the woman that she was very fair. The princes also of Pharaoh saw her, and commended her before Pharaoh: and the woman was taken into Pharaoh's house. He entreated Abram well for her sake: and he had sheep, and oxen, and he asses, and menservants, and maidservants, and she asses, and camels.'

Other positive human traits such as humour or wisdom that endear an individual to their peers or make them stand out in value above the crowd may also, in part, have been selected as survival traits in times of hardship such as famines. Unfortunately, as we have seen above, the same may be said for some of mankind's worst traits: selfishness, greed, the quest for power, a readiness to enslave our fellow man, to take away his land when he is destitute, to steal and to kill.

Societal and population survival strategies

As with the behavioural strategies there is much that has been written about the broader societal and population strategies that have been adopted through the ages to avoid famine. All of these involve elements of planning and organisation in order to create stores of food in times of plenty, or to move food from one region of a country to an afflicted region. Amartya Sen has noted that there has never been a famine in a functioning democracy, and that most famines can be traced to corrupt or inefficient practices at some level of government. But outside of democracies there have also been many governments that have created elaborate systems of taxation and tithes, of granaries and of food distribution when crops fail. Returning again to the Bible for reference to famines we recall Joseph's interpretation of the Pharaoh's dream of the seven kine [cows] coming out from the river 'fatfleshed and well favoured' and being consumed by the seven other kine 'poor and very ill favoured and leanfleshed, such as I never saw in all the land of Egypt for badness', and of the seven ears of corn 'full and good' that were devoured by the seven bad ears 'withered, thin, blasted by the east wind'. Joseph predicted seven years of plentiful harvests but that 'there shall arise after them seven years of famine; and all the plenty shall be forgotten in the land of Egypt; and the famine shall consume the land; And the plenty shall not be known in the land by reason of that famine following; for it shall be very grievous.' And the pharaoh was admonished to 'look out a man discreet and wise, and set him over the land of Egypt. Let Pharaoh do this, and let him appoint officers over the land, and take up the fifth part of the land of Egypt in the seven plenteous years.' This, in summary, has been the model for any successful system of famine avoidance be it in a democracy or a dictatorship.

But globalisation and rapid transport systems have created new possibilities whereby areas of the world struck by famine can, and should be, rescued by food imports or charitable aid. Almost without exception the only time such interventions fail is when governments wilfully obstruct their implementation as in Darfur, or Zimbabwe, or North Korea whose government was intent on denying that a problem existed.

For governments who have failed to care for their peoples famine has frequently been the tinderbox for uprisings and revolution. Zhu De, the Chinese Marshall (1886–1976) recalled such a threat from his childhood, in the second summer of a drought in Sichuan:

> From the dust cloud there soon emerged a mass of human skeletons, the men armed with every kind of weapon, foot-bound women carrying babies on their backs, and naked children with enormous stomachs and cavernous red eyes plodding wearily behind them...The avalanche of starving people poured down the Big Road, hundreds of them eddying into the Zhu courtyard, saying 'Come and eat of the big houses!'
>
> (Smedley 1956.)

In France, Marie-Antoinette's alleged statement *'Qu'ils mangent de la brioche'* aptly highlights the perils that may befall a complacent leadership. Simon Schama's *Citizens: A Chronical of the French Revolution* describes how, on 13 July 1788, a hailstorm burst over much of central France from Rouen as far south as Toulouse with 'stones so monstrous they killed hares and partridge and ripped branches off elm trees'. In the Île-de-France, where fruit and vegetable crops were wiped out, farmers wrote: 'A countryside erstwhile ravishing, has been reduced to an arid desert.' In much of France a drought followed, and that in turn was followed by a winter of a severity the like of which had not been seen since 1709, when the red Bordeaux was said to have frozen in Louis XIV's goblet. Frozen rivers prevented mills from turning what grain there was into flour and people were reduced to boiling tree bark to make gruel. The thaw bought further misery by flooding the fields of the Loire. Following a fact-finding tour of Provence Mirabeau reported that 'The region has been visited by the exterminating angel. Every scourge has been unloosed. Everywhere I have found men dead of cold and hunger.' And these natural events together with punitive taxes from the Fermière Générale were major contributors to an enraged peasantry finally storming the Bastille, and thus offering a memorable example of a form of societal response to surviving famine. It is a sad irony that Antoine Lavoisier, the founder of the study of energy metabolism that has allowed us to study the metabolic mechanisms for surviving famine, and a man who had done much to improve crop production and storage methods, was to lose his head by the guillotine in 1794.

Contemporary consequences of surviving famine

We have alluded already to the most obvious contemporary consequence of evolutionary selection by famine – namely the current pandemic of obesity and its associated diabetes (see Figure 7.11). It would seem a relatively uncontentious claim to argue that our metabolisms were designed to cope

FIGURE 7.11   Obesity. (Photo courtesy of Felicia Webb.)

best with periodic fluctuations between feast and famine, and are ill suited to cope with constant feasting. By such an argument the rapidly escalating worldwide rates of obesity are not a strange and perplexing quirk of metabolism, but are the absolutely predictable consequence of the mismatch between an ancient metabolism and a rapidly altered environment. It can further be argued that the human race has never before been challenged by such a rapid alteration in our ecological niche, and never before has the challenge been self-generated. The enormous health and economic consequences brewing on the horizon are a matter of great concern.

Much more contentious, but all the more interesting as a consequence, is the possibility that another metabolic and psychological illness, anorexia

FIGURE 7.12   Anorexia nervosa. (Photo courtesy of Felicia Webb.)

nervosa, may also be a maladaptive echo of a one-time adaptive trait that helped our forebears' struggle against famine (Figure 7.12). Guisinger (2003) has argued that some of the key pathological features of anorexia nervosa, namely an energetic restlessness and urge to move, searching for and an appreciation of the sight of food but with a suppression of hunger, enjoyment of feeding others, denial that the body is threatened by starvation, and an inappropriate optimism, could all represent survival traits. The reasoning is that when local resources are depleted migration to new feeding grounds is required. Optimism is required to believe that salvation is possible and to drive on the quest to move. A suppression of hunger is necessary to avoid the individual being trapped in the local area, or distracted en route, by the constant search for a morsel to eat, and yet it is important to maintain a longer-term food-seeking behaviour. It is further argued that just one or two individuals possessing such traits might rescue whole families or communities, as Queen Boadicea is reputed to have done. Although this theory may sound far-fetched, it resonates well with those who believe that there is a distant biological explanation to every human action and behaviour.

Malthusian predictions and the future

In 1798 Thomas Robert Malthus wrote his *Essay on the Principle of Population* in which he stated that:

> Population, when unchecked, increases in a geometrical ratio. Subsistence increases only in an arithmetical ratio. A slight acquaintance with numbers will shew the immensity of the first power in comparison of the second.

He went on to argue that:

> The power of population is so superior to the power of the earth to produce subsistence for man, that premature death must in some shape or other visit the human race. The vices of mankind are active and able ministers of depopulation. But should they fail in this war of extermination – sickly seasons, epidemics, pestilence, and plague, advance in terrific array, and sweep off their thousands and tens of thousands. Should success be still incomplete, gigantic inevitable famine stalks in the rear, and with one mighty blow levels the population with the food of the world.

Malthus anticipated that such a mighty famine would occur within a quarter of a century of his essay. It didn't. Others since then have been making similar predictions and continue to do so. As Malthus himself pointed out it is an argument in which the protagonists tend towards an unhelpful polarisation of their views. For example the modern harbingers of doom argue that world food production increased rapidly in the 1980s and 1990s but then plateaued, and interpret this as a worrying sign. They fail to admit that this is has been a consequence of overproduction that has depressed world grain prices and required massive investment in set-aside programmes to persuade rural farmers to reduce their output. They also frequently fail to account for the rapid slowing in the world's population growth rates.

A reasonable interpretation of current data is probably that we can feed ourselves into the future, but only so long as we avoid a global catastrophe. But can we do so? Recalling Ancel Keys' comment that man's dominance of nature allowing him to assume the role of 'creator of his own misery' one is forced to wonder whether humankind is not now colluding with nature and may yet validate Malthus' cataclysmic prediction. We will learn more in a later chapter, but one of the potential consequences of future climate change is a devastating reduction in crop yields in certain parts of the world. If this were to create a Malthusian imbalance between population and sustenance we will not be able to rely on an extraterrestrial NGO to drop food packages, nor will we be likely to be troubled by the difficult decision of whether to uproot our family and migrate to Mars. We have the technology to assess the situation – do we have the political will to act? Will a global famine be the exception to prove Amartya Sen's rule that famines do not occur in functioning democracies? Am I ending on too pessimistic a view – or, as Malthus said in the preface to his essay, with 'a jaundiced eye or an inherent spleen of disposition'? Perhaps so – but the warning signs are there for us to heed.

ACKNOWLEDGEMENTS

I am deeply grateful to photographers Tom Stoddart (Getty Images) and Felicia Webb for allowing me to use their remarkable pictures in my lecture and this chapter.

FURTHER READING

Alamgir, M. (1980). *Famine in South Asia*. Cambridge, MA: Delgeschlager, Gunn and Hain.

Becker, J. (1996). *Hungry Ghosts: China's Secret Famine*. London: John Murray.

Conquest, R. (2002). *The Harvest of Sorrow: Soviet Collectivisation and the Terror-Famine*. London: Pimlico.

Fagan, B. (2000). *Floods, Famines and Emperors*. London: Pimlico.

Jordan, W.C. (1996). *The Great Famine: Northern Europe in the Early Fourteenth Century*. Princeton: Princeton University Press.

Guisinger, S. (2003). 'Adapted to flee famine: adding an evolutionary perspective on anorexia nervosa', *Psychological Review* **110**, 745–61.

Kane, P. (1988). *Famine in China, 1959–61: Demographic and Social Implications*. London: Macmillan.

Keys, A., Brozek, J., Henschel, A., Mickelsen, O. and Taylor, H.L. (1950). *The Biology of Human Starvation*, 2 vols. Minneapolis: University of Minnesota Press.

Prentice, A.M. (2001). 'Fires of life: the struggles of an ancient metabolism in a modern world', *British Nutrition Foundation Bulletin* **26**, 13–27.

Sen, A. (1981). *Poverty and Famines: An Essay in Entitlement and Deprivation*. Oxford: Clarendon Press.

Woodham-Smith, C. (1991). *The Great Hunger: Ireland 1845–1849*. London: Penguin.

8 Surviving longer

CYNTHIA KENYON AND CLAIRE COCKCROFT[†]

Introduction

The previous few chapters have considered a range of threats to the survival of humankind, including threats to our longevity through the impacts of disease, natural disasters and famine. In this chapter I will continue an exploration of our longevity by considering how in the future we may be able to stay younger for longer. The biochemical processes I will describe may well have developed in part as an evolutionary response to past threats such as famine.

Genes from the fountain of youth

Finding the key to the universal process of ageing and unlocking secrets to the fountain of youth is one of the quests that humans have pursued for generations. I started to study ageing in the early 1990s, at a time when it was considered to be something that happened in a passive, haphazard way – an inevitable wearing out, like old cars – something that nothing much could be done about. How the rate of ageing is determined is not clear and far from simple. However, like all aspects of biology, the ageing process has now been shown to be under exquisite regulation by a complex, multifaceted hormonal system common to many species across the breadth of the animal kingdom.

[†] This chapter was written by Cynthia Kenyon and Claire Cockcroft, and is based on the Darwin Lecture given by Cynthia Kenyon about her research.

Survival, edited by Emily Shuckburgh. Published by Cambridge University Press.

I started to question the view that ageing was simply a passive and inevitable consequence of molecular wear and tear for several reasons. Firstly, looking around the natural world, it is immediately apparent that different animals display strikingly different lifespans: mice, for example, live for 2 years, canaries for about 15 years and bats can, rather surprisingly, live for up to 50 years. Rats live for just 3 years, but squirrels can live for 25. These animals differ from one another by their genes; therefore, genes must have an enormous influence on ageing.

Throughout the last twenty to thirty years of biological investigation, almost every process studied at the molecular level has turned out to be under very constant and sometimes highly elaborate genetic control. Often when scientists go off in some new, murky and potentially unfruitful area, it turns out to be fascinating voyage of discovery, revealing of all sorts of control. So, it seemed conceivable that something as universal as ageing might be subject to fundamental control. In other words, there may be *genes* that determine how rapidly we age. In this chapter I will describe how we have tested this in the laboratory by altering genes and observing the effect. The idea is that if there are genes that control the rate of ageing, then changing these genes might extend lifespan. Finding and studying such genes may reveal profound insights into the mechanisms governing ageing and potentially herald new approaches to disease management.

But rather than study humans directly, the focus of our mission to unravel the complexities of ageing has been the tiny, soil-dwelling nematode (see Figure 8.1) called *Caenorhabditis elegans*, the first multicellular organism to have its entire genome sequenced. These roundworms are microscopic creatures, just visible with the naked eye, and they are quite harmless. They reproduce rapidly and can be raised in large numbers in Petri dishes as shown in Figure 8.2, where they move gracefully on the surface of the bacterial lawn on which they graze, hence the name 'elegans'. The worms are transparent so that every cell in the living animal can be observed under the microscope from the fertilised egg to the adult worm, providing a valuable window into the world of animal development and a convenient model for biologists to study. They have a short lifespan, living only a few weeks, but before the worm dies it shows signs of ageing and thus may provide general clues about the ageing process.

FIGURE 8.1 *Caenorhabditis elegans*. Viewing the worms under a high-power microscope enables individual tissues to be observed, which can be seen to deteriorate with age.

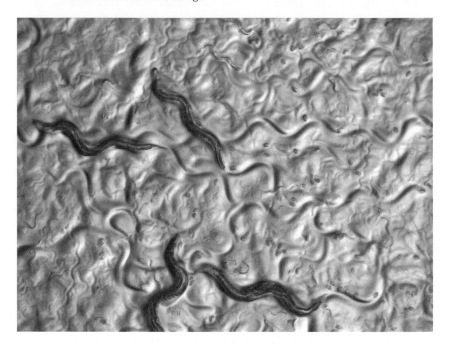

FIGURE 8.2 Young adult worms move elegantly on the bacterial field on which they feed, hence their name.

Although *C. elegans* is some evolutionary distance from humans, numerous biological processes have been conserved during evolution across the animal kingdom and it was therefore plausible that studying the genetic mechanisms governing ageing in one of the simplest multi-cellular animals may provide valuable insights into ageing in humans. For example, humans have nerve cells that make dopamine and *C. elegans* possesses six neurons that make dopamine as well as neurons that make serotonin and acetylcholine, illustrating conservation of neurotransmitters between species that have some evolutionary distance between them. The intestine and muscles also work in the same general way, so *C. elegans* is very much an animal in the most basic sense. Since ageing is something so fundamental to living things, we hoped that learning about its regulation in worms may have relevance in some way to people. By virtue of their short lifespan and ease of reproduction, *C. elegans* is an excellent model for the study of ageing.

The quest for long-lived mutants

The identification of 'long-lived mutants' in natural populations of *C. elegans* revealed that a number of genes do have a key role in influencing longevity. Early studies by David Friedman and Thomas Johnson had shown that altering one particular gene (*age-1*) elicited a 30–50% increase in lifespan. However, little was known about how this gene affected life span, and at this time, the field of ageing was considered something of a backwater by many molecular biologists. Finally, a student in my laboratory (Ramon Tabtiang) took up the challenge to find the key genes influencing ageing. The general idea was that if there really were genes affecting ageing, then one could identify them by looking for long-lived mutants. A population of *C. elegans* would be treated with an agent to damage its genes; in other words, a mutation would be made. We hoped that introducing mutations that extend lifespan would provide clues about the ageing process in the worm with the altered gene. The long-lived worms would be allowed to have descendants, which would be screened to find any with an enhanced lifespan and those long-lived offspring studied to see whether or not they also gave rise to long-lived progeny. This generational transmission would be expected if a gene involved with ageing had been affected. So we set out to look for long-lived mutant worms.

In 1993, we showed that mutating a single gene, called *daf-2*, in *C. elegans* doubled its lifespan (see Figure 8.3). The *daf-2* gene produces a protein called DAF-2. In the *daf-2* mutants the protein is slightly impaired; it has some function but not complete function. The normal *daf-2* gene

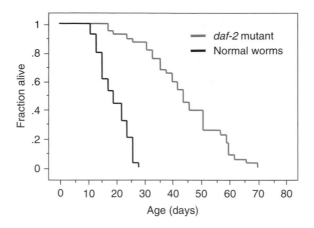

FIGURE 8.3 Variation in the fraction of worms alive with time. After one month all normal worms are dead whereas 90% of the *daf-2* mutant worms are still alive. After two months almost all the *daf-2* mutants are dead. Hence mutations that damage the *daf-2* gene double the worm's lifespan.

is therefore implicated in making the worm age because when we restrain its function in the mutants, the worms live longer – so the normal gene must be rather like a grim-reaper gene. Might humans also have a grim-reaper gene? I will come on to discuss this later, but first I will describe more worm studies.

Apart from their enhanced life span, there is no visible difference between normal (known as 'wild-type') and mutant worms. Normally in two weeks a wild-type population of worms will be growing old or dying, but the mutants are still very much alive and kicking, showing no indication of ageing in their tissues. These mutant worms appear to age more slowly than normal. Looking at one of these worms is comparable to looking at a 90-year-old person and thinking, from their appearance and level of activity, that they are only 45 years old. This is almost like a miracle, in that changing one gene can have such a striking effect, but it's not a miracle, it is robust, demonstrable science. The next stage was to try to understand how changing the *daf-2* gene can have such a dramatic effect on the worm.

The *daf-2* gene was already known before we implicated it in the ageing process. It was discovered by Sydney Brenner, the postdoctoral mentor of

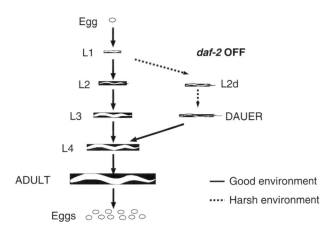

FIGURE 8.4   The life cycle of *C. elegans* (see text for details).

me of us (CK), over thirty years ago. Studying mutant worm populations had revealed that strong mutations in the *daf-2* gene (stronger than the ones in Figure 8.3, i.e. ones that further impared the protein DAF-2) had an effect on the animal as it developed from an egg to adult.

During the life cycle of *C. elegans* illustrated in Figure 8.4, when the worm hatches from the egg, it is very small in 'L1' but grows through the four different stages ('L1' to 'L4') and emerges an adult. This happens if there is plenty of food ('Good environment'), but if food is scarce ('Harsh environment') they do something rather clever: they don't die but exit the growth cycle during L1 and enter a state, rather like hibernation, called the *dauer* (derived from the German word for 'enduring'). The dauer resembles an L3, which is about the same age, but it is thinner and its growth is arrested – it doesn't eat, it doesn't reproduce, it just waits. Once it senses food, it exits from the dauer stage into an L4, from which it matures into a *normal* adult with a *normal* lifespan, even though it has taken 'time out'.

If, however, an adult is deprived of food, it does not become a dauer. Only animals that have not yet been through puberty can become dauers – there is a pre-puberty checkpoint. It is when the animal comes out of the dauer state that the reproductive system grows and matures, and the worm goes through puberty. The animal therefore assesses whether or not to mature and enter reproductive phase or, in the absence of a favourable nutritional situation, whether it should wait to produce its offspring until the unfavourable conditions have passed, and instead enter a phase of 'suspended animation'. The *daf-2* gene controls the decision to grow up or to become a dauer. If a mutation is made in *daf-2* that totally knocks out its function, essentially the gene is turned off, the animals hatch from eggs and become dauer larvae, irrespective of food availability. Thus, one normal function of the gene is to allow the animals to become adults. Mutants with partial activity of the gene, however, had enough activity to become adults and also live longer. The *daf-2* gene therefore has two roles – it controls the decision to be or not to be a dauer and also controls the life span of an adult.

Genetic influence on lifespan

At what point in the life cycle does the *daf-2* gene act to control lifespan? Is it when the animal is very young to control the future lifespan of the adult? Or does it act at the time the ageing process is actually taking place? Modern molecular tools enable us to 'turn down' the activity of a gene at any specific time in the life of the animal using a clever technique called *RNA interference* (or 'RNAi'). If *daf-2* activity is turned down throughout life, the worms live longer. If the worms grow up normally and are then subjected to RNAi treatment (i.e. *daf-2* activity is turned down only during adulthood) they also live longer – it is as if the gene had been turned down all through their entire life. We can experiment further to pinpoint the critical time for *daf-2* activity. Turning it down when the worm is developing into an adult and then turning it back up once it reaches adulthood results in worms with a normal lifespan. So if the gene is functional in the adult, the worms age normally. These experiments tell us *daf-2* functions exclusively during adulthood to affect lifespan.

The *daf-2* gene therefore has a role in two completely separate processes, 'dauer formation' and 'ageing.' Early in life, *daf-2* acts to decide if the animal will become a dauer or grow into an adult and then it acts separately during adulthood to control ageing. But how does *daf-2* orchestrate matters at the molecular level?

All genes, including those that control the physical development of the organism, are an intrinsic part of the DNA located in the 'nucleus' of a cell (cells are the building blocks of tissues). The information in the DNA is moved into RNA, and then used to make a protein. Proteins are the workhorses of the cell with a diverse range of functions and responsibilities. Knowing what is the function of the protein made by the *daf-2* gene may reveal mechanistic details about the ageing process. The *daf-2* gene was cloned in the laboratory of Gary Ruvkun at Harvard. He had been examining the process of dauer formation and the role of *daf-2* gene in the dauer pathway when we found that it affected ageing. He found that *daf-2* encodes a protein, known as DAF-2, which functions as a hormone receptor. This, together with our findings that mutations in *daf-2* double the worm's lifespan, allows us to come to the important conclusion that hormones control ageing (see Figure 8.5).

Stepping back from the science momentarily, people have been looking for the 'fountain of youth' for centuries, so we are excited by the prospect that the humble worm may help to unlock the secret of ageing. Our expectations were heightened when we learnt more about the *daf-2* gene and its role in other animals. The hormone receptor encoded by the gene is similar to two receptors encoded by human genes: one of these is a receptor for the hormone insulin and another for a hormone called insulin growth factor number 1, better known as IGF-1. The previous chapter discussed how insulin may broadly be linked to traits that enhance chances of survival in times of famine. Insulin is involved in the uptake of nutrients into the tissues of the body after eating a meal and IGF-1 is known to be involved in growth. Both are essential hormones, which, if completely knocked out in any animal from worms to humans, result in death.

Is it possible that these same hormones have another function in higher organisms including humans, namely to speed up ageing, as they do in worms? Although we don't know about humans yet, four different laboratories are investigating whether these same genes control ageing in other organisms and, remarkably, they do. Tatar and Partridge investigated the role of these genes and receptors in *Drosophila*, commonly known as 'fruit flies' and a model organism for biologists. If the gene encoding the insulin/IGF-1 receptor, the fly's version of *daf-2*, is damaged, the flies live about 80% longer. Similar results have been found with mice. Mice, like humans, have two separate genes encoding the receptors for insulin and IGF-1. Mice with a less active IGF-1 receptor were created in

(a)

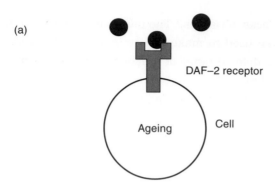

DAF−2 receptor

Ageing   Cell

(b) Hormones control ageing

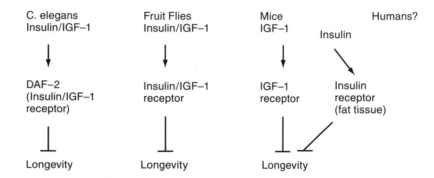

C. elegans Insulin/IGF−1	Fruit Flies Insulin/IGF−1	Mice IGF−1	Humans?
↓	↓	↓ Insulin ↘	
DAF−2 (Insulin/IGF−1 receptor)	Insulin/IGF−1 receptor	IGF−1 receptor	Insulin receptor (fat tissue)
⊥	⊥	⊥ ⁄	
Longevity	Longevity	Longevity	

FIGURE 8.5   (a) The *def-2* gene encodes a hormone receptor. A cell with a DAF-2 hormone receptor, represented as a sphere with a hormone-receptor baseball glove sticking through from the inside to the outside of the cell membrane (boundary). When the relevant hormone (solid spheres) is in the vicinity, the glove-like receptor grabs the hormone signal, rather like catching a baseball, and the cell does something new in response to the hormone. The normal function of the DAF-2 receptor is therefore, from its position in the cell membrane, to wait for a hormone signal in its environment and, after securing the hormone to the receptor, orchestrate a series of events within the cell that cause the cell to age more quickly. (b) Our work, together with findings from the Tatar, Partridge, Holzenberger and Kahn laboratories, has shown that hormones are implicated in ageing, in the worm and other animals, and that certain hormones speed up ageing.

the Holzenberger laboratory, and had increased lifespan. The mice were completely fertile, had normal respiration and lived about 20% longer. Ron Kahn's laboratory at Harvard showed that knocking out the insulin receptor not only increased the lifespan of the mice by about 20%, but the mice also remained thin if they were fed a high-fat diet, whereas a normal mouse would get fat.

This demonstrates evolutionary conservation of the mechanisms associated with ageing, confirming this is not specific to worms but that this hormone system affects ageing in several different animals, from worms and flies to higher animals like mice. These genes therefore acquired the ability to influence ageing a long time ago in evolution, before the divergence of worms, flies and mice from one another. The extent of this evolutionary conservation raises the possibility that these genes also control longevity in humans since we also arose from the common precursor of the worm, fly and mouse. It is a fundamental question, but we don't yet know if this is the case in humans: it is not so easy to find out by genetic manipulation like in the worm and fly, and even if we could, we would have to wait a long time to find out. However, it seems likely that, to some extent, these genes do control ageing in humans.

How do hormones ultimately influence how we age?

Hormones are tiny molecules circulating in the body, coming into contact with cells and tissues and eliciting a response to the stimulation. But it is not immediately obvious why one hormone circulating in the body would change the rate of ageing. From our wider knowledge of hormone function, the cellular response to hormones is often to switch certain genes on or off or to turn them up or down a notch. Although all the cells in the body have copies of the same DNA, not all of the genes are active in all of our cells, such as muscle, skin or intestinal cells – they can be either active or inactive. For example, in the heart, genes involved with muscle contraction are active, but those needed for vision are off. In the eye it's the opposite; genes that allow you to see are active but the genes allowing the heart to beat are off. We therefore wondered if changing the *daf-2* gene also changed the activities of other genes in the DNA besides the *daf-2* gene.

Could the DAF-2 receptor somehow control the activity of other genes in the same way that other hormone receptors control gene activities?

Today there are rather wonderful ways that allow scientists to profile all the genes – in the case of the worm 20 000 genes – to see how active they are. This technique, called 'microarray analysis' or 'micro-expression analysis', enables the profile of gene activity in a normal worm to be compared with the profile of gene activity in a long-lived mutant worm, thereby highlighting which 'downstream' genes are more or less active. We found that there were changes in expression activity of many genes: several hundred were either more active or less active in the long-lived worms. We did further experiments to explore whether these genes enhanced the life span, taking the top 50 genes identified earlier on the basis of how much their activity changed in the mutant worms. One by one we turned down each candidate gene using the technique of RNAi.

By taking a long-lived *daf-2* mutant worm and turning down the activities of these 'downstream' genes, one by one, we could see if the mutant worm still lived longer than normal worms. This revealed some interesting findings. Inhibiting the activity of many genes that are more active in long-lived *daf-2* mutants shortens the life of these mutants compared to those bearing only the *daf-2* mutation (see Figure 8.6). However, inhibiting the genes never reduced the life span below that of normal worms. In each case we got a fairly modest effect on life span, suggesting that the downstream genes act cumulatively to increase life span.

We then analysed the genes that are less active in the long-lived *daf-2* mutants. Conversely and somewhat surprisingly, some genes were less active in the long-lived animals. Genes that are less active would be genes that you would predict would speed up ageing because they are less active in an animal that has a long lifespan. We then turned these candidate genes down in a normal worm to see if that was enough to extend the lifespan of the worm. For many cases it was: turning down the activities of the genes that are less active in long-lived mutants increased the life span.

In summary both the downstream genes that are more active and the genes that are less active in long-lived mutants are important contributors to the ageing process. However, the effect that we get for altering any individual gene is not that large. Understanding exactly what all these genes are doing may help to illuminate the mechanisms governing ageing.

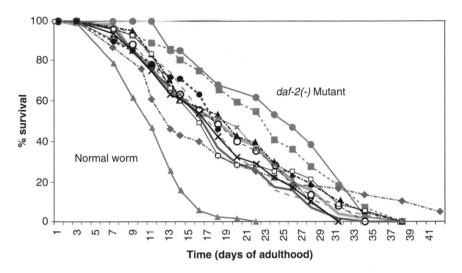

FIGURE 8.6 The effect of inhibiting genes that are more active in the long-lived *daf-2* mutants. The ageing of a *daf-2* mutant (squares) is delayed compared to a normal worm (triangles). The other lines are also for *daf-2* mutant worms but they have had an additional single gene turned down that is normally more active in long-lived mutants. Inhibiting the activities of these genes individually prevents the worm from living as long as it normally would and therefore these genes must also contribute to the long life span of *daf-2* mutants. (Murphy, C. T. *et al.* (2003). 'Genes that act downstream of DAF-16 to influence the lifespan of *Caenorhabditis elegans*', *Nature* **424**, 277–83.)

As the DNA sequences of all 20 000 genes in *C. elegans* are known, we can find out the sequences of the particular genes we are interested in. Taking this sequence information, looking at databases to find out what proteins the DNA encodes and what proteins they resemble from other species has told us a lot about the mechanism of ageing.

It is clear that these genes do not have just one role; they underpin many different processes in worm life. For example, there are antioxidant genes, which prevent reactive oxygen species from flourishing and causing cellular damage in the animal. These genes encode proteins like superoxide dismutase, catalase and glutathione S-transferase, which have an antioxidant capacity. Antioxidants are found in many foods, like blueberries, which nutritionists encourage us to consume to stave off the consequences of cellular damage caused by free radicals. These antioxidant genes are more active in the long-lived animals, which has also been demonstrated for a few genes by several laboratories previously, and now, from our study, appear to enhance longevity of the *daf-2* mutants.

There are also genes with a beautifully descriptive name – 'chaperones'. As the name implies, a chaperone protein takes care of other proteins, helping them to assemble correctly and fold into the proper shape so that they can function effectively. If a protein is badly damaged, the chaperone will escort it to the equivalent of the cell's rubbish bin for destruction. In the long-lived mutant worms, the chaperone genes were found to be more active, so there were more chaperones in the animals, also potentially important contributors to their long lifespan. Gordon Lithgow's laboratory introduced extra copies of a chaperone gene into normal worms and showed that they lived longer, again arguing that these genes are implicated in enhanced longevity.

We also found that the worm's immune system had been recruited, as several genes that make proteins to destroy bacteria and fungi were more active in the long-lived animals. Worms frequently die from infections, particularly as they age. If we add an antibiotic to the culture dish in which we grow the worms, so the bacteria are alive but cannot divide, then the worms live about 30% longer. They seem to get frail at the same rate, but under conditions normally experienced in the wild, as worms get increasingly frail, bacteria can more easily puncture them and invade their bodies, causing an infection and ultimately death. Long-lived animals live much longer than just 30% longer: they live twice as long, but we think at least one reason is that they have all these antimicrobial genes working at higher rates, thereby contributing to the long lifespan of the worms. There are of course parallels with humans as infections are a significant cause of illness and death in humans. For many previous generations, infections were often the cause of death before ageing processes had manifested themselves.

A number of metabolic genes were also affected the in *daf-2* mutants with increased lifespan. Apolipoprotein genes make special proteins that carry fat around the body from place to place. We don't know why these genes affect lifespan, but they are less active in the long-lived *daf-2* mutant worms and, if we make them less active in a normal worm, the worm lives longer, implying that they have an important role. Interestingly, there are certain genes that have been shown to be associated with people who live to 100 years of age. This ability to become a centenarian runs in families and, studying centenarian siblings, scientists are attempting to map these genes to understand what they do and how they enable people reach 100 years. In three out of three cases so far, genes involved in fat transport have been pinpointed, and one of them is the human version of the genes that we found affected the worm. In centenarians they seem to be turned down, just as they are in the long-lived worms. So, although the population size is small, it is a nice link between what we are finding in worms and human populations.

Other laboratories have found other individual genes that were changed in the long-lived mutants thereby potentially contributing to increased longevity. The big picture from all this work is that ageing is not caused by just one factor. There are many different ways of combating ageing or slowing it down in these worms, from antioxidants, chaperones, antimicrobial proteins, metabolic genes to genes that we don't yet understand called 'novel' genes. One could think of it like an orchestra, where different instruments such as flutes, violins, cellos and so forth (here the different genes) are all tied together by the conductor (here the *daf-2* gene), which makes them play harmoniously in concert to orchestrate effects on lifespan.

This is an important finding with numerous implications. Firstly this tells us that there is a genetic control module governing life span, with a director of operations, the DAF-2 receptor, at the top, influencing many subordinate or downstream genes. But how could a regulatory module like this arise during evolution? It controls ageing but it is possible that it evolved actually because it allowed animals to age at certain rates?

There is another explanation of how this gene network could have evolved. In addition to controlling ageing, the *daf-2* gene also controls entry into the dauer state, the hibernation-like state induced by pending starvation that is only accessible before the animal goes through puberty. Some of the genes affecting lifespan also affect the dauer. These dauers are very long-lived and are also resistant to all sorts of noxious treatments: if you douse them in hydrogen peroxide they survive; if you heat them up they survive, so they're highly resistant to diverse kinds of environmental stress. And proteins like the antioxidant proteins and the chaperones exist to protect an animal against these types of stress. It is therefore possible that this regulatory mechanism arose as an escape clause so that an animal on its way to adulthood can 'opt out' if it senses nutritionally challenging or famine-like conditions ahead, thereby safeguarding its survival until a time of plenty returns. So you can see how there would be a Darwinian selective pressure for something like this to arise during evolution because an animal that is smart enough, from a teleological perspective, to wait until environmental conditions improve before it has all its progeny, will have an advantage over one that simply matures and bears progeny at a time when their survival prospects are low.

It could be that this control mechanism evolved for a very practical reason, for the dauer state, enabling the animal to have time out, to survive famine. But once present in the worm's genetic make-up, it can also be used to control the ageing process because the downstream genes that protect the dauer from environmental stress can also be used protect the animal from their inevitable fate of death by the ageing process. Knowing the relationships that these genes have to one another, and to the DAF-2 receptor, it is likely that the same genes conferring protection against environmental stress may also be able to protect the animal against metabolic stress, for example the oxidative damage generated simply through the process of being alive, or against proteins that misfold and have the potential to cause age-related diseases. Consider the ageing paradoxes between similar species: the rats who live for 3 years and squirrels who live for 25. Could changes in regulators like *daf-2* or other downstream genes be responsible for the differences in lifespan between different species? This can be tested in the laboratory now that we know what the genes are.

Ageing and onset of disease

This brings me to another interesting and important question, namely what links the normal ageing process to age-related disease? Humans are 100 times more likely to get a tumour at the age of 65 than at 35. But it is not simply because you have lived for a specific number of years – mice, for example, get old and die at around 2 years, developing cancer very frequently at about 1½ years and dogs are prone to cancer at about 10 years. It is being elderly that makes the individual susceptible to cancer and other diseases, rather than that a specific number of years or days have elapsed. But what is the link?

The insulin/IGF-1 hormone system appears to link ageing to many age-related diseases. Scientists have looked at a range of diseases in the long-lived worm mutants. One laboratory has put the protein causing Huntington's disease into worms – which resulted in the worms getting the equivalent of Huntington's disease as they age. However, in long-lived *daf-2* mutants, the response is delayed, as though they are protected from the toxic effects of the disease-causing protein. Normal worms develop a muscle condition during ageing that resembles a human muscle-wasting condition known as sarcopenia, although long-lived mutants do not develop this condition until they are much older. We have also shown that long-lived *C. elegans* mutants are resistant to tumours.

In fruit flies it transpires that the insulin/IGF-1 hormone system controls the ageing of the heart. If you take a normal fly when it's young and stress the heart electrically, nothing happens, the fly is fine. But if you give the same stress to an old fly, the heart will fail. However, if long-lived *daf-2*-equivalent mutant flies are subjected to this stress, their hearts never fail: even when the flies are on their last legs, the heart is in better shape than the rest of the animal. So this gene, and the hormone response it controls, really has an effect on cardiac function.

Interesting results also have been discovered in mice. Long-lived mutants have been shown to be resistant to cancer as well as other diseases and every few months there are new reports about other diseases that are postponed in one of these long-lived animals. This is an interesting revelation – that these long-lived animals age more slowly in the most fundamental ways – because diseases of ageing don't afflict them until later.

Much is known about genes controlling ageing and we have identified around fifty genes that are key candidates. The fact that there is a link between ageing and age-related diseases raises the possibility of a potentially powerful therapeutic strategy. If we could develop a pill that slowed ageing, it might also have some effectiveness against treating a whole variety of age-related diseases. It's an entirely different approach to the treatment of age-related diseases. The laboratory of Rick Morimoto in Chicago has developed a model of Huntington's disease in the worm by putting the human gene that causes Huntington's disease into the worm. We have used these worms see which of the genes associated with increased life span are linked with the ability of the worm to resist the onset of Huntington's disease.

Figure 8.7 shows the *C. elegans* model of Huntington's disease. The worm is glowing on the sides due to a fluorescently tagged Huntington's protein. The Morimoto laboratory put the gene into the worm in such a way that it is active in the muscle cells so the bright dots illuminate the muscles of the worm containing this fluorescent protein. The disease is caused by an altered form of the Huntington gene producing a faulty Huntington's protein, which forms aggregates – clumping together leading to illness and ultimately death. In young worms the protein hasn't aggregated, but as it gets older the protein starts to clump, seen as the little dots in Figure 8.7 (bottom), and as the worms age they become paralysed. As in humans with Huntington's disease, the Huntington's protein aggregates with increasing age in the worm. Furthermore, the paralysis caused by the Huntington protein is delayed in long-lived *daf-2* mutants, so they retain the ability to move around a lot longer than would normally be the case, even though they have the faulty Huntington's gene.

Huntington's disease protein forms aggregates in worms as they age, as in humans

The paralysis caused by Huntington's protein is delayed in a long-lived *daf-2* mutant

FIGURE 8.7 Fluorescence micrographs illustrating the accumulation of fluorescently-tagged Huntington's protein aggregates in *C. elegans*.

The question is which of the genes that we identified as acting down-stream from *daf-2* could encode the molecular links between ageing and this paralysis? There must be something else beyond the hormone itself circulating in the blood that elicits this effect. From the list of suspects, the chaperones caught our attention because they are known to prevent unfolded or damaged proteins from clumping together – precisely the issue in Huntington's disease. So we thought these chaperones might couple ageing to the onset of such age-related disease. To investigate this we used the RNAi technique to see the effect of inhibiting the chaperones in worms with the faulty Huntington's gene. Turning down the chaperone genes caused the protein to aggregate more quickly, in younger animals, demonstrating a link between ageing and this aspect of Huntington's disease.

Acknowledging that this is not the whole story, it appears that the regulation of chaperone genes is one way that the hormone system can couple normal ageing to the timing of disease development. We hope that this knowledge will be applicable to other diseases as laboratories around the world try to understand what exactly it is about ageing that makes one susceptible to a disease.

Can environmental cues affect this ageing system?

We were interested to explore whether there might be some environmental condition that changes the level of the hormones and then changes lifespan? Hormones are present under some conditions but not others – for example human stress hormones are low if you are relaxed but rise rapidly during a stressful encounter. We think there is an equivalent in the worm – remarkably. The sense of smell and taste can affect the levels of worm hormones and consequently the activity of this receptor and life span of the animal. In other words we think the lifespan of *C. elegans* may be regulated by sensory perception.

Several mutant worms have been created with defects in different 'sensory genes', which normally allow the worms to smell and taste. We discovered that these sensory mutants all live longer. Some of these genes make proteins called chemosensory receptor proteins, which are located at the tip of the nostrils. One side of the protein is embedded in the cell and the other is outside enabling the worm to physically respond to compounds in the environment that it smells or tastes. Changing the genes that make those chemosensory proteins extends the worm's life.

These changes did not alter the worm's feeding behaviour – they eat normally – hence we think that it is sensory perception itself that affects lifespan. More specifically, our experiments suggest that there is some environmental signal that the worm senses through its sensory neurons, causing the worms to release more of the insulin-like hormones (see Figure 8.8). We already know that the genes that make the hormones are active in these neurons, so we hypothesise that, under normal circumstances in a laboratory environment, the hormone is released, it binds to the DAF-2 receptor on the surface of the cell and the worm has a normal lifespan. But if we disrupt this chain of events by meddling with neuron development, the worms release less hormone, rather like having an empty receptor, so the worms live longer.

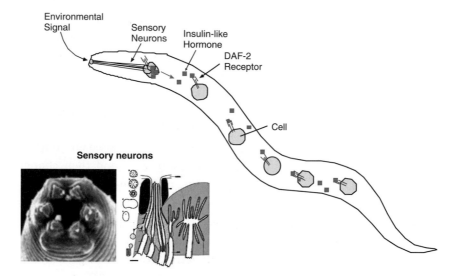

FIGURE 8.8 Proposed model of sensory perception in *C. elegans*. The lifespan of *C. elegans* is influenced by its perception of soluble and volatile substances in the environment. The insets illustrate a close-up of *C. elegans'* face with the nostrils clearly visible (left) and these nostrils in finer detail (right) showing the individual sensory neurons, which allow worms not only to smell but taste things with their nose too. Mutations can be made in the genes that produce the neurons, giving rise to worms with poor smell and taste perception.

More detailed experiments were performed to dissect this process further. Each nostril contains twelve neurons, which we knew were responsible because if the whole nostril was destroyed then the worms lived longer. But which of the twelve pairs of neurons was responsible? Some of the neurons allow animals to smell while others allow the animals to taste, so we started destroying them with a laser, one by one, to discover which neurons influence life span. Destroying one gustatory (taste) neuron extends the lifespan, which is amazing. Surprisingly we found that if we destroyed the same taste neuron plus also another neuron, called 'J' for short, they don't live longer. Therefore, some neurons promote long life and others inhibit long life: they are not all the same. In summary, if we destroy one particular taste neuron, they live long. If we destroy that same taste neuron plus the J neuron they have a normal lifespan. What about smelling though?

We can also kill neurons called olfactory neurons that allow the worm to smell and destroying one particular smell neuron extends the lifespan. If we destroy both smell and also the taste neurons in the same animal they live even longer. But what happens if you destroy the smell neuron plus the J neuron, do you get a normal lifespan? It turns out we still get a long lifespan. The J neuron counteracts the taste neuron but does not have any effect on the smell neuron. So the J neuron can 'communicate' with certain neurons, such as the taste neurons, but not with the smell neurons. We don't know how elaborate the communication gets between neurons but from this and other experiments it appears that the lifespan of *C. elegans* is influenced by environmental signals, such as volatile substances that it can smell and those that it can taste. We don't know what it is yet but I suspect it is something that may tell the worm that an environmentally unfavourable situation is around the corner.

Now for the critical question – could this be relevant to humans? In fruit flies, unpublished findings indicate that if you prevent a fruit fly from smelling it also lives longer. If a human eats a meal the level of the hormone insulin in the blood rises, but if you also smell the food you're eating it goes up even more. Insulin is the hormone that has been implicated in the control of lifespan, which isn't to say that it is the same in humans, but it might be possible.

Is there a reproductive trade-off?

Evolutionary biologists are fascinated by the relationship between ageing and reproduction. Is there some kind of a reproductive trade-off? In the mutant worms, we can change one gene and they live longer, but are they less able to reproduce than normal? Many people think there has to be a trade-off, in that for every incremental increase in lifespan something deleterious must happen. However, during evolution, the animal lifespan increased from three weeks in the case of the worm to the long lifespan of humans today – more than a thousand-fold increase. So that means that if every time there was a mutation in an ancestral gene that increased lifespan there was always some downside, we would be pretty miserable creatures. In other words, it is difficult to see how humans could have evolved to such a degree of sophistication. We can alter the *daf-2* gene in worms in such a way that they live twice as long as normal but reproduce normally. On the other hand if you damage the gene a little bit more severely, the worms live longer but produce fewer progeny, so there can be a trade-off, but it isn't necessary.

But then we found something really interesting, that the *daf-2* gene acts in worms at different times to control ageing and reproduction. Using the RNAi technique, we discovered that in the adult *daf-2* controls ageing, but it acts earlier to control reproduction: if we use RNAi to turn down *daf-2* activity throughout the worm's life, it lives twice as long as normal and has slightly delayed reproduction. But if the worms grew up with a normal *daf-2* gene and then we turned *daf-2* down on day one only of adulthood, they lived twice as long as normal, reproduced on a normal schedule and produced the same number of progeny at the same time as normal worms. This is an interesting observation telling us that no trade-off is necessary between reproduction and ageing, at least under these conditions, and implies that changes in the *daf-2* gene during evolution could potentially change both lifespan and reproductive timing. However, it is conceivable that changing *daf-2* in a different way may provide a means of affecting one process but not the other, since there is not a causal relationship between them. It is not that one is a consequence of the other – producing fewer progeny because you live longer or vice versa – instead the same gene controls both ageing and reproduction and they can be separated from one another as we did in the laboratory, or they can be co-expressed or co-active.

I would now like to ask whether cause and effect should be entirely reversed. In other words, do the cells in the reproductive system actually control the lifespan of the animal? If you look at an animal when it's a newly hatched worm, it has only four cells in the reproductive system (see Figure 8.9a). Two cells give rise to the germline – the sperm and the oocyte (white) and two more cells (grey) give rise to what we call the somatic gonad, the reproductive tissues like the uterus and the spermatheca and so forth. We discovered that if precursors to the germline cells are destroyed using a laser (see Figure 8.9b), so that the animals have no sperm or oocytes, they live 60 % longer than normal, a dramatic effect. However, these animals are sterile, which one may think explains why they live longer. Experiments have shown that this is not the case: if we destroy all four cells using a laser so they have no reproductive system whatsoever, they have a normal lifespan but are sterile (see Figure 8.9c). So it

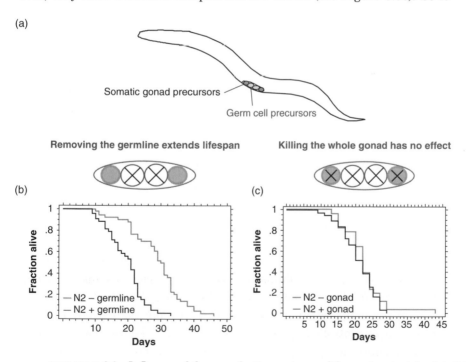

FIGURE 8.9 Influence of the reproductive system on lifespan (see text for details).

is not just a trade-off between fertility and ageing, it must be something different.

Instead, the germ cells and also the somatic cells seem to be doing something actively that affects the lifespan of the animal. The germ cells control a hormone that affects ageing, a steroid hormone. Insulin is a small protein hormone but a steroid hormone is chemically similar to something like testosterone or oestrogen, a small organic molecule with lots of ring structures in it. Why would this kind of control process have evolved? The answer is not clear, but we can imagine the following: one of the important things that all animals have to do is co-ordinate ageing with the timing of reproduction. All animals reproduce best when they are in their prime so perhaps this system provides that kind of co-ordination. If there were a random mutation in a worm causing its germ cells to develop more slowly, thereby taking longer for the animal to have progeny, this would delay both the germ cell development and the birth of progeny. One might wonder, is the worm then going to be too old to have progeny? We argue that it wouldn't be because, as a consequence of slowing the rate of ageing, the animal would still potentially be in its prime when it reproduces. This illustrates how a system like this could bring these two important processes – ageing and reproduction – into alignment with one another.

This led us to ask the ultimate really cool question of how long could a worm live? We don't actually know the answer to that but we discovered that, if the *daf-2* gene is altered in a worm inhibiting its activity, and we also alter the reproductive system, the worms live *six times* longer than normal. The worms live, on average, 126 days and are healthy, whereas normal worms in the same experiment only live for 20 days. Two worms lived to 144 days old, and still looked perfectly healthy. From their general appearance and levels of physical activity, they were judged more akin to worms aged 5 days, which is just amazing.

This pales in significance compared to the 1000-fold increase in lifespan that has arisen during evolution, but the fact that making two small alterations can elicit such an enhanced lifespan is remarkable.

So, in summary, ageing really doesn't 'just happen': it is not a passive process, it is controlled by numerous hormones, themselves affected by environmental signals, or the germ cells. In response to the hormone activity, the DAF-2 receptor can control the activities of many more genes co-ordinating different cellular events within the animal to influence life span.

Since the theme of this book is 'survival', just for fun, I thought I would raise some interesting food-for-thought questions about ageing and extending lifespan in humans. Is it possible that we could stay young for longer, ethical implications aside? Changes during evolution have already increased our lifespans enormously, several thousand-fold, but have we reached the upper limit? If we consider the germ lineage, which produces the sperm and oocyte, this is immortal – it started at the beginning of life and is still going strong. So if this cell lineage can go on forever, is our lifespan completely cast in stone? Or, consistent with our findings and those of other laboratories, perhaps one can think about the ageing process as a competition between the forces of damage and entropy which tend to age the cells, counteracted by the forces of repair and prevention, mediated through antioxidants and chaperones for instance. One could imagine that the average lifespan of a species of animal is determined by the ratio of these two factors: for example a mouse with a short lifespan has a much higher rate of damage than prevention, whereas in humans the ratio would be somewhat lower. In other words, in humans, the ratio of prevention and repair to damage is higher, enabling us to live much longer than mice. Imagine if a cell could boost the protection and repair mechanisms so that they completely counterbalance cellular damage – is it possible that a cell might not age? It may not be possible but it is a thought-provoking scenario.

Humans already have a long lifespan but why do we live as long as we do, way beyond the time needed to produce offspring and even to get them out of the house? If we look first at the worm, which has a very short lifespan, its ultimate strategy is to turn any source of food into more worms as soon as possible. It grows into an adult in 3 days and in the next three days has 300 offspring, which grow up in 3 days. So by the time a worm is 6 days old it has 300 brand new worms, each of which is capable of producing 300 worms, compounding at the level of 300% every 3 days. Humans do not have that strategy or capability to compound that fast; our strategy is to organise. We thrive because we organise, control resources and deploy them for the benefit of the tribe, which takes time and is generally better served by a longer lifespan.

So why do we have these long post-reproductive lifespans? Two possible reasons have been suggested that I agree with. One reason why we may live so long is that the elders are valuable, which still holds true today – those running institutions, businesses and countries for example are not 20-year-olds, they are often in their sixties. This indicates that there must be value in longevity because the 20-year-olds are much stronger and could easily change things if it were to their advantage to do so. Ageing and maturity has long been associated with enhanced wisdom. British judges still wear grey wigs, and indeed in past times wigs were actually worn by young people often with a full head of dark hair, as numerous portraits hanging in Cambridge colleges attest. Wigs made people look older and more dignified, and gave them a sense of power. So in many ways people would like to be older because of the perceived value and wisdom associated with age. Moreover one can see a clear benefit by looking at tribal existence, before there was the written word. If the only source of knowledge is that retained by the elders, and some form of disease or natural disaster recurs that is only remembered by the elders, then the elders impart a clear advantage to the tribe. In fact, the death rate in the recent tsunami in Indonesia was much lower among families with grandparents, who knew, from past experience, what to do when the waters recede. In addition, it is possible that we have such long post-reproductive lifespans in order to help to take care of our grandchildren, thereby giving their parents (our children) more time to hunt or engage in other activities that benefit the the family.

Now we can ask, given that it is valuable that we have long lifespans, why is it that we don't live even longer? Is it because we have reached a limit? I would suggest that perhaps there isn't a lot of selective value in living much longer. If the grandparents are around to help care for the grandchildren, maybe the great-grandparents are no longer needed. And perhaps it isn't necessary to remember back more than sixty years.

Together all of these arguments provide a nice explanation for why we live as long as we do and not longer. We don't need to postulate that we don't live longer because it's simply impossible – we have reached the limit. If we could stay young and disease-free for longer, then why haven't scientists been hard at work on this problem for a long time? I think it's because we didn't have any role models. We invented aeroplanes because we saw, and envied, birds. There are animals that live longer than we do, but, people generally thought, who wants to be a sea turtle? Now that we can see these little animals, like worms and mice, staying young and living longer, we are getting the idea that it just might be possible for people. They, these little long-lived mutants, are the new role models.

In conclusion, from our early discoveries in the 1990s, revealing a twofold enhancement in longevity, we have manipulated other genes and cells and have now been able to extend the lifespan and period of youthfulness of *C. elegans* sixfold. We have found that signals from the reproductive system and sensory neurons influence life span and that these signals act, at least in part, to control insulin/IGF-1 and steroid hormone signalling. These hormones influence the life span of the animal by co-ordinating the expression of a wide variety of subordinate genes, including antioxidant, stress response, antimicrobial and novel genes, whose activities act in a cumulative fashion to determine the life span of the animal. Some of these genes can also influence the rate of onset of age-related diseases, such as tumour formation. In this way, the hormone system couples the natural ageing process to age-related disease susceptibility.

Unravelling the universal control mechanisms that have evolved to safeguard survival in lower animals is providing insights into the ways environmental signals are interpreted by cells to protect against the inevitable processes of wear and tear as the animal ages. Our work and the work of the many other laboratories studying ageing in the worm, together with findings in other higher animals, indicates that there are definite prospects for novel therapeutic and lifestyle approaches for humans, which may ultimately allow us in the future to stay young longer.

FURTHER READING

Brown, A. (2003). *In the Beginning Was the Worm*. London: Simon & Shuster.

Complete issue of the journal *Cell* on ageing: *Cell* (2005) **120**, 435–567.

Friedman, D.B. and Johnson, T. E. (1988). 'A mutation in the *age-1* gene in *Caenorhabditis elegans* lengthens life and reduces hermaphrodite fertility', *Genetics* **118**, 75–86.

Hekimi, S. (ed.) (2000). *The Molecular Genetics of Aging: Results and Problems in Cell Differentiation*. New York: Springer.

Kenyon, C. (1997). 'Environmental factors and gene activities that influence lifespan', in Riddle, D. L., Blumenthal, T., Meyer, B. J. and Priess, J. R. (eds.) *C. elegans II*. New York: Cold Spring Harbor Laboratory Press.

Kenyon, C. (2005). 'The plasticity of aging: Insights from long-lived mutants', *Cell* **120**, 449–60.

Murphy, C.T., McCarroll, S.A., Bargmann, C.I. *et al.* (2003). 'Genes that act downstream of DAF-16 to influence the lifespan of *Caenorhabditis elegans*', *Nature* **424**, 277–83.

Strauss, E. (2001). 'Longevity: growing old together', *Science* **292**, 41–3.

Tatar, M., Bartke, A. and Antebi, A. (2003). 'The endocrine regulation of aging by insulin-like signals', *Science* **299**, 1346–51.

9 Survival into the future

DIANA LIVERMAN

Introduction

This book started with a discussion of the Social Darwinists and their belief that the key to survival in the late nineteenth century world was to *organise*. Here in the final chapter of the book I will discuss the urgent need for *global organisation* to address the new threat posed by global climate change.

Survival into the future in the face of climate change

Climate change presents one of the great challenges for the survival of people and ecosystems in and beyond this century. Not for humanity or the planet as a whole, but for significant numbers of vulnerable people and ecosystems.

The Arctic as we know it, with its wildlife and indigenous peoples, will not survive a warming climate. The Alps are already losing the winter snow cover that supports tourism and the Andes are losing the snow and glaciers that feed water resources. Sea-level rise will threaten low-lying islands, coastal development and ecosystems, and drought will add to the problems of the poor in Africa, unless we take serious action to reduce the concentration of greenhouse gases in the atmosphere and to adapt livelihoods and landscapes to a warmer climate.

The last couple of years have brought a sea change in knowledge and management of climate risks with new research showing that global

Survival, edited by Emily Shuckburgh. Published by Cambridge University Press.
© Darwin College 2008.

warming is now with us and that the risks are greater than previously estimated. In 2005, the Kyoto Protocol came into force and the European Union (EU) began trade in the new commodity of 'carbon credits'. At the climate negotiations in Montreal in 2005 and Nairobi in 2006 issues of development and adaptation moved forcefully up the international agenda. It is on these new developments that this chapter will focus, in order to add something to the climate change discussion with which we are all becoming familiar.

I will first review of some of the most recent research on the risks of climate change. Then I will present an evaluation of the adequacy of UK and international response to these risks in terms of reducing greenhouse gas emissions. Finally, I will make the case for an expanded focus on adaptation to climate change.

Understanding these issues requires unprecedented levels of scientific and political collaboration – between natural and social scientists and between governments, corporations and individuals – collaborations that in themselves pose great challenges of communication and governance. This chapter draws heavily on the interdisciplinary research at the Environmental Change Institute at the University of Oxford, in the Tyndall Centre and in the Intergovernmental Panel on Climate Change (IPCC).

Understanding the risks

New insights provided by science into climate change risks to the survival of people and ecosystems include the message that climate change is here and the probability of severe and rapid change and impacts is increasing.

Most people are familiar with the basics of the climate change problem whereby human activities such as fossil fuel consumption have increased the concentration of greenhouse gases in the atmosphere since 1850 by over 50% to 2005 (from 278 parts per million by volume (ppmv) carbon dioxide equivalent to around 425 ppmv in 2005). International scientific assessments have suggested that should the concentration double to 550 ppmv the average global temperature would increase by 3 °C, with a greater increase at the poles.[1] The European Union has adopted 2 °C as its target for stabilising climate and avoiding dangerous change.

Research has now accumulated that shows not only that the risks of climate change at 550 ppmv may be more severe and more discontinuous than previously assumed and that dangerous changes may occur below a 2 °C threshold, but also that the 50% increase in greenhouse gas concentrations to date is already producing observable changes in climate.

Robust scientific analysis, summarised in graphics from the United Kingdom's Hadley Centre has demonstrated that global temperatures have increased by 0.75 °C, sea level has risen about 15 cm, glaciers have

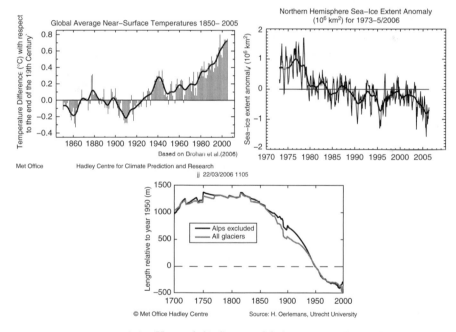

FIGURE 9.1 (Upper left) Strong global warming observed since 1975; (upper right) Arctic sea-ice extent decrease since the mid 1970s; (lower) Glaciers have retreated globally. (Hadley Centre (2005). *Climate Change and the Greenhouse Effect.* © Crown copyright 2005, published by the Meterological office. avaliable from www.metoffice.gov.uk.)

[1] The latest IPCC report (2007) indicates that by 2099, depending on economic development scenarios, temperature will have risen by 1.1 to 6.4 °C.

retreated and Arctic sea ice has contracted (Figure 9.1)[2]. These changes are starting to affect ecosystems as fragmenting ice cover disrupts polar bear and caribou. Warmer temperatures change the seasonal and spatial behaviour of birds and plants in Europe. Human lives are disturbed as salt water intrudes into wells on low-lying islands, and the frequency of high-intensity storms increases.

Shifts in our understanding of future climate risks have emerged from work on the *climateprediction.net* experiment. Thousands of computer simulations are being used to explore uncertainties in climate prediction. The experiment is run on personal computers around the world in a fascinating public participation exercise recently expanded in partnership with the BBC. The results suggest a shift in the probability distribution of climate risk with the slight but worrying possibility of warming of up to 11 °C if greenhouse gases reach 550 ppmv.

The new climate science is summarised in the recent book on *Avoiding Dangerous Climate Change* based on a conference held in Exeter in 2005. The conference provided important insights into climate risks especially the risks of delaying cuts in emissions and of large-scale discontinuities. The conference theme was driven by the wording of the United Nations Framework Convention on Climate Change which has the goal of 'stabilizing greenhouse gas concentrations in the atmosphere at a level that would prevent dangerous anthropogenic interference with the climate system'. This simple statement hides a host of scientific and political challenges, not the least of which is to define and establish a threshold for what is 'dangerous' and then to link this to a stable level of greenhouse gases.

Stabilisation targets have included '550 ppmv', '450 ppmv' and 'a 2 °C warming' adopted formally by the EU. The Exeter papers show that even these targets may increase the risks and bring forward the timing of large-scale and sudden changes. More probable and proximate changes discussed in Exeter and in the forthcoming IPCC reports include the collapse of the Antarctic, the reversal of flows of heat and salt water in the North Atlantic, the release of methane from northern latitudes, and the disruption of cycles of the monsoon and El Niño.

---

[2] Similar changes are documented in the 2007 IPCC report, which also notes an increase in extreme rainfall events, severe droughts and intense hurricanes.

The critical point for the melting of the Greenland ice cap would be a temperature increase of 2.7 °C. Temperature rises above 3 °C increase the probability of the destabilisation of the Antarctic and a reversal of the ability of land and vegetation to act as a carbon sink in regions such as the Amazon basin.

Even if we stabilised greenhouse gas concentrations at current levels of about 400 ppmv we will have committed the world to further climate change because the gases are long-lived and even current warming has some feedbacks which may influence future climate and concentrations. Because of uncertainties in the climate system we could easily overshoot even a 2 °C target. We are also starting to understand that delaying reductions by only ten years will produce a much more serious warming and require a larger cut in emissions than if we act now.

What we are beginning to realise is that targets such as 'a maximum 2 °C increase' or 'stabilised emission concentrations' may not be the best way to design climate policy because of the scientific uncertainties and challenges in maintaining a stable emission level. It may be more scientifically and strategically robust to focus on policies that eventually bring emissions back down towards pre-industrial concentrations – essentially bringing the carbon cycle back into its historical balance – and to manage both mitigation and adaptation so as to constrain peak warming and its impacts.

Some species and cultures may not survive

New scientific observations and predictions have also changed the way we are thinking about the impacts of climate change and provide a link to previous chapters in this book.

The image that is often used to synthesise ideas about the impacts of dangerous climate change is known as the 'burning embers' diagram (Figure 9.2). This diagram links average global temperature changes to risks to a set of five concerns – unique and threatened ecosystems, extreme climate events, distribution of impacts, overall impacts and large-scale discontinuities. Reading across to the 'embers' (columns) the lower range of temperatures would equate to moderate risks to some systems and of extreme climate events but low risk of large

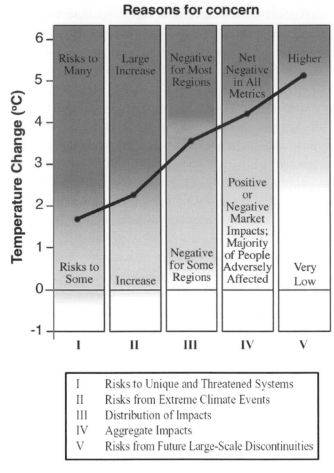

FIGURE 9.2   The burning embers (note original in colour with white–grey–black scale being yellow–orange–red). (Mastrandrea, M. D. and Schneider, S. H. (2004). 'Probabilistic integrated assessment of "dangerous" climate change', *Science* **304**, 571–4.)

aggregate economic impacts and of large-scale discontinuities. The 5 °C change brings high risks to many systems, of extremes, of unequal and high aggregate economic impacts, and a moderate risk of large-scale discontinuities.

It is at 5 °C that we glimpse what might be considered risks to survival of species and cultures. We start to move into the range of the large-scale discontinuities and this is likely to exacerbate existing threats discussed in previous chapters on disease, natural disasters, famine and so on. Five degrees lies well within the range of possibilities for climate change if we do not achieve significant reductions in greenhouse gases soon.

Biophysically the most vulnerable regions are the polar zones, mountain tops, low-lying islands and coasts and it is here that we can see immediate survival threats.

In the Arctic, which has already seen warming in some regions of 4 °C, a further 6 °C warming is seen as possible by the Arctic Climate Impact Assessment. This would mean the widespread encroachment of boreal forest into the tundra, the disappearance of sea ice, the loss of habitat for ice-dependent polar bears and seals, and stressing caribou and reindeer populations. The loss of the traditional landscape and ecology will drastically alter the culture and economy of Arctic peoples.

Ecosystems are already responding predictably to the warming and the survival of a number of species is increasingly at risk. Research suggests that a large percentage of species are changing their phenology, moving their distribution polewards and upwards, or changing their overall population numbers. As temperatures warm, species that live in cooler highlands are being squeezed to higher altitudes, bringing them into conflict with other species and risking extinction as their habitat shrinks and disappears from the drier or warmer mountain top (Table 9.1). One effort to assess extinction risk suggested that 15% to 37% of all species over a representative 20% of the world's land area could be driven extinct from the climate change that is likely to occur between now and 2050. This led to the widely quoted 'climate change threatens a million species with extinction'.

For low-lying island nations such as in the Pacific, 5 °C, especially if accompanied by Antarctic melting, could mean the abandonment of their viable territory through sea-level rise, saline intrusion into water resources and higher risks from tropical cyclones. A global assessment of risks to survival reported at Exeter suggests millions more people at risk from coastal flooding, food shortages, malaria and hunger at a 2 °C or more temperature rise.

Table 9.1 *A globally coherent fingerprint of climate impacts across natural systems*

Change	Studied	Significant change	As expected
Phenological (timing)			
Woody plants	38	31	30
Herbaceous plants	38	12	12
Mixed plants	385	325	279
Birds	168	92	78
Insects	35	13	13
Amphibians	12	9	9
Fish	2	2	2
Total	678	484	423 (87%)
Distributional			
Polewards/upper			81%
Equatorwards/lower			75%
Abundance			
Cold-adapted			74%
Warm-adapted			91%
Total	920	460	372 (81%)
Meta-analyses			
Phenologies	172	Earlier spring timing of 2.3 days per decade	
Range boundaries	99	6.1 km per decade northward or per decade upward	

Source: Parmesan and Yohe (2003).

Double exposure

Biophysical vulnerability is only one element in surviving climate change, where impacts often depend on the social and economic characteristics of affected peoples and on the intersection of climate change with other pressures on ecosystems. The risks of sea-level rise and severe storms are vastly compounded when millions of poor people live along low-lying coasts and will have to adapt, migrate or perish en masse as a result of climate change. The risks of species extinction are higher when land use prevents a shift in range or harvesting places extra pressure on populations at risk. Even more complex are the interactions between climate risks and other stresses to our food supply where millions may be at risk from changing climate, land degradation, and poverty in Africa.

My own work in Mexico has shown how climate change risks to agriculture are extremely difficult to define because everything else in the agricultural landscape is changing as a result of land reform, privatisation and free trade. Farmers in Mexico and other parts of the developing world are faced with 'double exposure' to the risks of climate change and of economic restructuring that has left them without agricultural extension, in debt, with declining prices for their products and paying high costs for water and fertiliser. Many of these farmers are deciding they cannot survive as agricultural producers and are moving to the cities or to the United States in search of alternative livelihoods.

Climate change poses real risks to the survival of ecosystems and cultures which are compounded by other changes that are taking place in the world economy. These risks are increasing the pressures to respond to climate change by cutting greenhouse gas emissions. There was a sigh of relief when at last, with Russian ratification, the Kyoto Protocol went into force with commitments to reduce greenhouse gas emissions and the risks of climate change.

Kyoto is too little but hopefully not too late

The entry into force of the Kyoto Protocol in 2005 formalised a set of international responses to climate change focused on reducing emissions of greenhouse gases through a combination of binding commitments on industrialised nations, a carbon credit trading system and development mechanisms that provide credit for investing in greenhouse gas reductions in the developing world.

Corporations, local governments and individuals are also seeking to reduce emissions through programmes that include energy conservation, renewables and carbon offsets. Dozens of new companies and consultancies have emerged to service the new carbon economy.

These responses are inadequate to the magnitude of the threats posed by climate change, especially if the sensitivity of the climate to greenhouse gases proves to be at the higher end of current predictions. The first stage of the Kyoto process is likely to produce an insignificant reduction and even an increase in greenhouse gas emissions, because of loopholes in the agreements and the absence of the United States and developing

Table 9.2 *Change in carbon emissions (million metric tonnes)*

Commitments and loopholes	Reduction	Cumulative
Toronto Conference 1988: 20% of 1988 emissions by 2005	−1334	
Kyoto Protocol 1997: 5.2% of 1990 emissions from industrialised countries	−770 (+564)	
Varying baseline year for some countries and gases	+53	+53
Including sinks in baseline	+34	+87
Carbon trading by 2012	+150	+237
Clean Development Mechanism by 2012	+292	+529
Credit for forest and land management	+171	+700
Excluding aviation and marine emissions	+90	+790
US non-participation by 2012	+520	+1310
Missing targets (EU) by 2012	+170	+1480
Developing world since 1990	+3000	+4480

Source: Based on data from Carbon Dioxide Information Analysis Center (http://cdiac.ornl.gov); Den Elzen, M. and De Moor, A. P. G. (2001). *The Bonn Agreement and Marrakesh Accords: An Updated Analysis* (www.rivm.nl/bibliotheek/ rapporten/728001017.pdf); Hare, B. (2000). *Undermining the Kyoto Protocol: Environmental Effectiveness versus Political Expediency* (http://archive.greenpeace.org/ climate/politics/reports/loopholes.pdf); UN Framework Convention on Climate Change (http://unfccc.int); World Resources Institute Climate Analysis Indicators Tool (http://cait.wri.org).

countries, and given that several other nations are unable to meet their commitments. The decision to base the first stage of Kyoto on carbon emissions in 1990 and to establish a market in carbon have created an arbitrary and unequal regime that is unlikely to achieve the reductions that are needed to prevent dangerous climate change.

Table 9.2 shows the progressive weakening of international commitment to reduce greenhouse gas emissions since the benchmark Toronto Conference on the Changing Atmosphere which in 1988 proposed a 20% cut in emissions from a 1988 baseline. This would have brought down worldwide emissions by at least 1.33 gigatonnes (Gt).

When it came to the concrete commitments at Kyoto in 1997, equity concerns and differentiated responsibility meant that the first phase of the Kyoto Protocol would only involve the industrialised countries and that considerable variation was allowed between them. The target cut was an average of 5.2% of a 1990 baseline by 2012, equating to 0.77 Gt (an increase of 0.56 Gt above the Toronto targets). This baseline was set somewhat arbitrarily and was clearly to the advantage of the United Kingdom and Germany whose emissions dropped significantly in the early 1990s for other reasons. In the United Kingdom, Margaret Thatcher initiated a move from coal to gas partly to break the miners' union and the collapse of the East German economy produced a drop in emissions for a unified Germany. Japan, in contrast, has a tougher time meeting their 6% target reduction because their economy already had high levels of efficiency making domestic reductions more difficult to achieve. The 5.2% target was politically feasible rather than scientifically meaningful and gave only about half the reduction proposed at Toronto.

The overall environmental impact was weakened yet again when some countries were permitted to use base years other than 1990, including Bulgaria and Hungary. Others were allowed to include emissions from deforestation in their base year emission total giving Japan, Canada and Russia a more beneficial position with higher initial emissions from which to make their assigned cuts.

Allowing countries to buy carbon credits rather than make more expensive domestic reductions reduced the benefits that had accrued from the collapse of former Soviet block economies whose emissions had declined far below their Kyoto targets and could now sell their unused quota through carbon trading.

Allowing countries to obtain credit for what I will show to be rather questionable investments in developing countries through Kyoto's Clean Development Mechanism further reduced the environmental significance of Kyoto, as did allowing countries to count carbon sequestering land management towards their commitments.

Leaving aviation and marine emissions out of Kyoto implementation was a further loss. The exit of the United States from the agreement (and an increase in its emissions) made things much worse.

And unfortunately reports from the UN and EU suggest that many of the countries within Kyoto are unlikely to meet their commitments. Although the United Kingdom, Germany, Russia and most of the eastern European countries have been meeting their Kyoto commitments a number of countries are a long way from meeting their targets. Between 1990 and 2003 Austria increased emissions by 16.5% rather than moving towards an agreed cut of 8%, Spain increased emissions by 41.7% (rather than an agreed increase of 15%) and Canada increased by 24.2% in contrast to a 6% target cut.

And, meanwhile, emissions in the developing world, with no obligations in the current climate regime have soared. Rather than the hoped for *cut* of 1.3 Gt in Toronto and 0.7 Gt in Kyoto we are looking at an *increase* of as much as 4 Gt by 2012 – as can be seen in Table 9.2.

The flexible but flawed Kyoto mechanisms

Understanding the problems of the international climate regime demands a closer look at Kyoto's flexible but flawed mechanisms for reducing the cost of emission reductions. Countries could (and have in some cases) met their Kyoto commitments purely through a combination of domestic standards, incentives and regulations ranging from fuel efficiency and building standards to incentives for fuel-switching and new technologies.

The introduction of flexible mechanisms involving carbon trading and credit for investments in carbon mitigation overseas effectively assigned property rights to pollute the atmospheric commons based on levels of pollution at the 1990 baseline. These mechanisms have allowed countries to obtain credit towards their Kyoto obligations at a much cheaper cost than reducing carbon emissions domestically. But for all their ingenuity these mechanisms do not get the world any further towards lower overall emissions. And, according to some critics, many of the flexible options do not produce verifiable reductions. The first real year of their operation (since Kyoto came into force in February 2005) revealed a number of flaws.

Carbon trading

The first Kyoto mechanism is that of carbon trading between industrialised countries. Advocates of market solutions see the commodification of the atmosphere as the most efficient way to reduce emissions using lowest-cost solutions. Critics see it as one more example of accumulation by dispossession in which the atmosphere – a common property of humanity – is divided up among nations based on arbitrary and sometimes advantageous conditions in 1990. It can be thought of as the privatisation of the atmosphere through an act of enclosure similar to the way non-owned land has been turned into private property around the world.

Windfall profits are made by those who were easily or unintentionally able to reduce emissions below target and then sell their excess credits, and by those who profited from the transaction costs of the new carbon economy. Russia whose emissions dropped 38.5% from 1990 to 2003 against a zero target is currently sitting on carbon credits worth $20–60 billion. Allocating rights to the atmosphere based on 'grandfathered' emissions in 1990 rewarded those with higher pollution (especially if they then had the political power to negotiate a small or shared percentage reduction).

It also set a dangerous precedent for a future round of negotiations with developing countries who might see an increase in emissions before the baseline as a good idea. Without trading Kyoto would have had a stronger environmental impact because countries would have been required to reduce domestic emissions rather than buy credits overseas. This might have produced more aggressive efforts to begin the restructuring of economies and innovations in technology towards lower carbon futures.

The only fair basis for trading, critics argue, is to give each person an equal share in the form of a personal carbon allowance. It would now be very difficult to reverse the carbon market because billions of dollars and many interests and actors are now tied up in its operation.

Since Kyoto came into force in 2005 carbon trading has taken off especially within the 25-nation EU Greenhouse Gas Emission Trading Scheme (ETS). The cost of carbon within the ETS in 2005 varied between €25 and €30 per tonne. The ETS has created a carbon market which according to the World Bank and to Point Carbon traded 362 million tonnes of carbon dioxide between 11 500 energy intensive installations across Europe worth €7.2 billion in 2005 with forecasts of almost €23 billion for 2006.

Critics of carbon trading point to the difficulties of monitoring and enforcement. Countries inventory their own emissions based mainly on national economic activity data and the data on land use change is particularly difficult to track despite international efforts to create a common methodology. In 2006, the price of carbon in the ETS dropped suddenly when it was revealed that many industries had been set generous emission quotas which could be easily met and exceeded – generating an oversupply of carbon reduction credits in relation to a lower demand.

The Clean Development Mechanism and sustainable development

It is another mechanism – the Clean Development Mechanism (CDM) – that has attracted the largest emission reduction credits of 397 million tonnes in 2005. The CDM provides credits for emission reduction projects in developing countries. The CDM was designed to help industrialised countries meet their emission reductions more cheaply and to contribute to sustainable development in the developing world through economic and technology transfer. So far it is working well as a cheaper way for countries to buy their way to their Kyoto commitments but is failing as sustainable development.

Problems with the CDM are partly associated with the inclusion of six greenhouse gases in the scheme and with the treatment of forests. CDM credits can be granted for five gases in addition to carbon dioxide (CO_2), including methane (CH_4) (waste disposal, fossil fuels, livestock), nitrous oxide (N_2O) (mainly from acid manufacture), perfluorocarbons (PFC) (from aluminium and semiconductor production), hydrofluorocarbons (HFC) (from refrigerant and polymer production) and sulphur hexafluoride (SF_6) (from electrical switchgear and magnesium production). All of these gases have a much higher global warming potential than carbon dioxide. Over a 100-year period a tonne of CH_4 is 21 times more effective in warming the atmosphere than CO_2, N_2O 310 times and PFC, HFC and SF_6 more than 1000 times higher.

What this means is that a project to reduce any one of the five other gases will generate a much higher number of emission reduction certificates, often at a lower cost, than one focused on the equivalent savings of carbon dioxide.

It is also important to note that at present the CDM can only be used for projects that replant forests (reforestation) or create new ones (afforestation), not for efforts to avoid the destruction of existing forests.

As of March 2006 there were 127 registered projects proposing 260 million tonnes of carbon reductions by 2012. Korea, Brazil, China and India host 88% of the carbon reductions, with the majority of the reductions associated with HFC destruction or nitrous oxide projects at six single industrial facilities.

Critics of the CDM are concerned about this bias to large industrial facilities and to larger developing countries with the easy reductions that they generate for their partners taking up demand that might otherwise promote sustainable energy and poverty reduction in the least developed world.

A more serious concern is that many CDM projects have provided spurious greenhouse gas reductions because they are associated with new developments (where the CDM might prevent future emissions but do not help the world return to 1990 or pre-industrial levels). It seems that many projects were already under way and the emission reductions would have occurred anyway through 'business as usual' and do not provide additional reductions as a result of the CDM financing.

The CDM originally had a strong 'additionality' requirement to prove that emission reductions would not have occurred without the CDM. But some of the initial CDM projects show only a difference in emissions with and without the project (meeting what is called an environmental additionality requirement). A wind project proposal will propose a coal plant as the alternative hypothetical scenario and shows that it has lower emissions and is thus environmentally additional whereas the true spirit of CDM was that the project should show that it was actually replacing the coal plant as a result of the CDM. If CDM projects are mostly derived from business as usual projects then credits they generate will not provide real reductions in emissions. And the inclusion of 'business as usual' projects, critics argue, has kept the cost of CDM credits down (€5–10 per tonne) and made it more difficult for costlier new renewable projects to obtain funding.

Other concerns relate to the expense of the CDM for smaller projects and poorer countries. Transaction costs for project development, registration and monitoring can reach $200 000 for even smaller projects. Some countries have been unable to benefit because of the difficulties in setting up and staffing the required institutions. Others are concerned that the system is managed by a World Bank agency – the Global Environment Facility – and thus reflects a governance system in which the rich countries control decisions.

Efforts have been made within Europe and at the Montreal climate negotiations to respond to some of these concerns. The EU will moderately constrain emission reductions through trade or the CDM in order to promote domestic cuts. The CDM has been modified to provide credits at lower transaction costs for small-scale renewables.

Plans are under way to consider clusters of related projects (e.g. for a city or an energy sector) as well as credits for overall policies that reduce emissions (sectoral CDM). There are proposals by Papua New Guinea and Costa Rica to include forest protection as avoided deforestation credits within the CDM. These discussions are critical to the next stage of Kyoto negotiations which must involve larger cuts and the involvement of the developing world after 2012.

Beyond Kyoto

It is clear that avoiding dangerous climate change is going to require more than tinkering with the Kyoto Agreement and requires much deeper and sustained cuts in concentrations of atmospheric greenhouse gases. While the UK government has cut greenhouse gas emissions by almost 15% (their share of the EU Kyoto target was 12.5%) it is off track in terms of meeting a domestic commitment to a 20% reduction in carbon dioxide by 2010 and a 60% reduction in carbon emissions by 2050.

'Surviving climate change' means a total rethinking of environmental governance and the restructuring of the economy towards a lower carbon future in both the developed and the developing world.

The route to deeper cuts has been demonstrated by projects such as the '40% house' (conducted by ECI and the Tyndall Centre) that shows how the UK housing sector could cut carbon emissions by 60% through energy-efficiency standards and new design for housing and appliances, demolition or refurbishment of poor quality housing stock, local combined heat and power (CHP) and renewable energy.

A vision for global major emission reductions over the next fifty years is provided by Princeton's 'wedges' that each cut one gigatonne of carbon through improvements in energy efficiency and conservation, shifts from coal to gas, carbon capture and storage, renewables, forest management and nuclear power (Table 9.3).

It is doubtful that market mechanisms or nation–states alone can deliver these cuts so we must consider a much wider range of policies and actors if we are to make a difference. Commitments by non-nation–state actors have

Table 9.3 *How to reduce 1 Gt of carbon in 2050*

Category	Subcategory	Method
Fuel shifting to displace coal	Electric plants	1400 GW fuelled by gas instead of coal
Carbon capture and storage	CO_2 stored in power plants	700 1-GW coal plants
	Geological sequestration	3500 like Norwegian prototype
	Hydrogen automotive fuel	1 billion H_2 cars displace 1 billion 30-mpg gasoline/diesel
Increased energy efficiency	Overall	7% of 2050 'expected' fossil fuel extraction
	Vehicles only	2 billion cars at 60 mpg instead of 30 mpg
Renewables for electricity	Wind displaces coal	70 × current
	Solar PV displaces coal	1000 × current; 5 × 106 ha
	Nuclear displaces coal	700 1-GW plants (1.5 × current)
Substitution of petroleum	Biomass fuels	200 × 106 ha, growing at 7.5 t(C)/ha-yr
	Hydrogen using nuclear energy	600 1-GW plants
Biological sequestration	Forests and agricultural soils	700 × 106 ha, growing at 2 t(C)/ha-yr

Source: Pacala and Socolow (2004).

increased dramatically in recent months and are being lauded a solution to the lack of US government participation and the weaknesses of Kyoto. Major corporations, including energy companies, have set out agendas for emission reductions and are moving to higher standards in response to investor concerns and to gain competitive advantage. Cities and regions have established climate change mitigation policies and thousands of individuals decided to compensate for their personal household and travel emissions through voluntary carbon offset programmes.

It is no easy matter to move away from fossil fuels because of the investment in infrastructure – cars, power plants, houses – that will not turn over for many years and because countries such as China will rely on coal to fuel economic development for years to come. Even carbon capture and storage needs a decade to secure technology and public acceptance.

So far we know very little about the concrete significance of the commitments of non-nation–state actors in terms of the overall level of greenhouse gases in the atmosphere. It is unlikely that they will achieve rapid and large reductions, even in combination with an international regime such as Kyoto, without a mix of high and stable carbon prices, long-term government policy or business agreements that set strict goals and standards, new technologies and a set of incentives for lower carbon economies in the developing world. In summary, it is probably that the world is committed to some serious level of climate change.

The urgency of adaptation

If climate change is occurring and current mitigation policies are inadequate, more attention must be paid to adaptation, and in particular, how the most vulnerable regions and people might find ways to cope with climate change. What type of *adaptations* are needed and who should finance them?

Many human societies have adapted to environmental extremes over the centuries – developing irrigation to cope with dry conditions and dams to control floods, terracing hillsides to prevent erosion and frost damage to crops, breeding crops and animals that are resistant to drought and disease, and engineering buildings to protect from high and low temperatures. Other species also adapt to changes in their environment, through, for example, evolutionary change or migration. While we can reinvigorate these traditional strategies, the rate, nature and scope of adaptation needed for climate change is unprecedented, especially if the climate change is experienced through extreme events or sudden thresholds.

In the Arctic, adaptation may mean widespread changes in traditional activities, the rebuilding of infrastructure. On small low-lying islands in the Pacific and Caribbean, adaptations include reinforcing coastal defences and warning systems, redrilling wells or providing alternative water sources, and finding new employment opportunities. Where millions of people are at risk from coastal inundation or crop failures adaptations will probably include a rethinking of disaster management and of food security.

There are some areas where adaptation can occur to reap the benefits of climate change for the story is not relentlessly negative. In the Arctic water transport will become easier, and regions will become favourable to agriculture and forestry. But planning for adaptation is hampered by continued uncertainty about regional climate impacts – in many regions we do not know whether to plan for more or less rain, and assessments of agricultural opportunity are complicated by other changes in the agricultural system. We also need to consider that some adaptations have unintended ecological and social consequences such as the effects of dams on downstream ecosystems and fisheries.

The Tyndall Centre has highlighted the critical issues of equity and of capacity in adapting to climate change. Decades of research on disasters has shown that the poorest groups in society are often the most vulnerable – forced to live in hazardous locations without the incomes, institutions and political power to cope with climate extremes. Many of the countries with the greatest urgency to adapt are those who bear the least responsibility for the changes and lack the resources to fund needed adaptations.

How will these countries find the funds for adaptation? Within international agreements such as the United Nations Framework Convention on Climate Change and agencies such as UNEP and the World Bank funds for adaptation are scarce. There are two voluntary funds and one that is financed through a tax on the CDM. These are completely inadequate to the task at hand. Aviation emissions are currently outside the Kyoto protocol and the ETS and some people have proposed that a tax on aviation fuel might provide a global fund for adaptation.

Another more controversial alternative is emerging from the science of attributing climate events and damages to global warming. It is possible to show statistically that anthropogenic warming, that is the warming attributable to our carbon emissions, doubled the risk of the European heat wave of 2003. As climate changes it will become even easier to link emissions to damages and possible for victims to sue major emitters for liability and thus secure a source of funding for adaptation.

In Europe we are already investing billions in quiet adaptations to a changing climate as we rebuild infrastructure after floods, install air conditioning in new buildings, buy snow-making machines, reinforce sea walls, and switch crops. A recent study estimated adaptation costs as high as €31 billion for coastal protection, €73 billion in construction costs and €64 billion in lost tourism revenues. We need to raise the profile of adaptation and consider how to 'mainstream' adaptation to climate change into everyone's plans and decisions.

We need to rethink our approach to nature conservation and landscape preservation as the optimum climate for native species shifts northwards. We need to rethink foreign assistance to include adaptation in our policies towards poorer countries and to support international efforts to establish a just framework for adaptation and the funds to implement it. We must readjust our management of water resources to cope with possibilities of drier summers and winter flooding. And, wherever possible, we need to find options that combine mitigation and adaptation such as designing buildings that produce less emissions but are also cooler in summer.

FURTHER READING

Allen, M. R. (2003). 'Liability for climate change', *Nature* **421**, 891–2.

Arctic Climate Impact Assessment (2004). *Impacts of a Warming Arctic*. Cambridge: Cambridge University Press.

Environmental Change Institute (2005). *The 40% House*. Oxford: Oxford University Press. (www.40percent.org.uk or http://www.eci.ox.ac.uk)

Hadley Centre (2005). *Climate Change and the Greenhouse Effect* (www.metoffice.com/research/hadleycentre/pubs/brochures/2005/climategreenhouse.pdf)

IPCC (Intergovermental Panel on Climate Change) (2007). *Fourth Assessment Report (Working Groups I, II and III)*. (www.ipcc.ch)

Pacala, S. and Socolow, R. (2004). 'Stabilization wedges: solving the climate problem for the next 50 years with current technologies', *Science* **305**, 968–72.

Parmesan, C. and Yohe, G. (2003). 'A globally coherent fingerprint of climate change impacts across natural systems', *Nature* **421**, 37–42.

Schellnhuber, H. J., Cramer, W. *et al.* (eds.) (2006). *Avoiding Dangerous Climate Change*. Cambridge: Cambridge University Press.

Tyndall Centre for Climate Change (2007). (www.tyndall.ac.uk)

World Bank (2006). *State of the Carbon Market*. Washington, DC: World Bank.

Epilogue

In all aspects of human life, political, economic and social, battles for survival rage. Indeed they have been an integral part of the history of *Homo sapiens* over the past hundred thousand years. The chapters of this book have examined just a few of the struggles, old and new. Looking to the future one thing is clear, that as a species we cannot become complacent if, in Charles Darwin's words, we are to 'look with some confidence to a secure future of equally inappreciable length', for the Struggle for Survival never ends.

Notes on the contributors

Emily Shuckburgh is a Fellow of Darwin College, Cambridge and a Research Fellow at the British Antarctic Survey. She is a mathematician who studies the physics of the atmosphere and oceans, considering environmental problems such as ozone depletion and climate change. She is a co-founder of the Institute of Aviation and the Environment at the University of Cambridge. She has an active involvement with the media, regularly contributing to discussions of climate change and other environmental issues on radio and television.

Paul Kennedy is the J. Richardson Dilworth Professor of History and Director of International Security Studies at Yale University, and internationally known for his writings and commentaries on global political, economic and strategic issues. His best-known work is *The Rise and Fall of the Great Powers*, which provoked an immense debate. His recent books include *Preparing for the Twenty-First Century* and *The Parliament of Man: The Past, Present and Future of the United Nations*, and a large collections of papers relating to contemporary strategic issues he co-edited entitled *From War to Peace: Altered Strategic Landscapes in the Twentieth Century*.

Edith Hall is Research Professor in Classics and in Drama and Theatre at Royal Holloway, University of London (and was previously Leverhulme Chair of Greek Cultural History at the University of Durham 2001–06). She is the co-founder and Co-Director of the Archive of Performances of Greek and Roman Drama at Oxford. Her books include *Inventing the*

Barbarian: Greek Self-Definition through Tragedy, an edition of Aeschylus' *Persians*, *Greek and Roman Actors* which she co-edited and *Greek Tragedy and the British Stage 1660–1914* of which she is a co-author. She regularly writes in *The Times Literary Supplement* and reviews theatre productions on radio.

Peter Austin holds the Märit Rausing Chair in Field Linguistics at the School of Oriental and African Studies in London and is Director of the Endangered Languages Academic Programme. A linguist, he is an expert on Australian Aboriginal languages and has conducted extensive field-work with Aboriginal communities. He has co-authored the first fully page-formatted hypertext dictionary on the World Wide Web, a bilingual dictionary of Gamilaraay (Kamilaroi), as well as publishing seven bilingual dictionaries of Aboriginal languages. He has also carried out research on Sasak and Sumbawan, Austronesian languages spoken on Lombok and Sumbawa Islands, eastern Indonesia.

Richard Feachem was the first Executive Director of the Global Fund to Fight AIDS, Tuberculosis and Malaria, and Under Secretary-General of the United Nations. He is Professor of International Health at the University of California–San Francisco and the University of California–Berkeley, and the founding Director of the Institute for Global Health at the two institutions. He was previously Director for Health, Nutrition and Population at the World Bank and Dean of the London School of Hygiene and Tropical Medicine. He has worked in international health and development for thirty years and has published extensively on public health and health policy.

Oliver Sabot is the Director of Medical Programs at the Clinton Foundation HIV/AIDS Initiative. He previously served as the Executive Communications Officer at the Global Fund to Fight AIDS, Tuberculosis and Malaria and is a founding member of Friends of the Global Fight, a Washington DC-based advocacy organisation. He is the author of several scholarly publications on global health and malaria in the *Lancet* and the *Journal of the American Medical Association*, among others.

James Jackson is Professor of Active Tectonics in the Department of Earth Sciences, University of Cambridge and a Fellow of the Royal Society. A geologist with a particular interest in earthquakes, he has done field-work in many parts of Asia, the Mediterranean, Africa, New Zealand and North America. He studies the evolution and deformation of the continents on all scales, from the movement of individual faults in earthquakes to the evolution of mountain belts, using space-based remote sensing combined with observations of the landscape in the field. In 1995 he delivered the Royal Institution/BBC Christmas Lectures on *Planet Earth: An Explorer's Guide*.

Andrew Prentice is Professor of International Nutrition at the London School of Hygiene and Tropical Medicine and a Fellow of Darwin College. He is also the Scientific Director of the Medical Research Council's Nutrition Programme based around the rural village of Keneba in The Gambia, West Africa. He also has collaborative projects in Pakistan, Bangladesh, Chile, Kenya, Zanzibar and South Africa. His research has considered many aspects of nutrition from the effects of malnutrition to the causes of obesity. He has a strong interest in the evolutionary consequences of famine particularly as mediated through effects on human reproduction. His work has been recognised by a number of international awards.

Cynthia Kenyon is an American Cancer Society Professor at the University of California–San Francisco and Director of the Hillblom Center for the Biology of Aging. In 1993 her discovery that a single-gene mutation could double the lifespan of *Caenorhabditis elegans* sparked an intensive study of the molecular biology of ageing. Her findings, for which she has received many honours, have now led to the discovery that an evolutionarily conserved hormone signalling system controls ageing in other organisms as well, including mammals. She is a member of the US National Academy of Sciences, the American Academy of Arts and Sciences and the Institute of Medicine.

Claire Cockcroft is the Deputy of Corporate Affairs at the Babraham Institute, managing external communications for the Babraham Research Campus and a Science and Society Programme. Previously she founded and co-directed the M.Phil. in bioscience enterprise at the University of

Cambridge, after completing doctoral and postdoctoral studies in plant molecular biology and biotechnology. She has written a number of popular-science articles, including for the *Guardian* newspaper, and now focuses on science communication and science public engagement activities.

Diana Liverman is Professor of Environmental Science and Director of the Environmental Change Institute at Oxford University. She has chaired the US National Academy of Sciences Committee on the Human Dimensions of Global Environmental Change, and has sat on advisory committees for NOAA, NASA and the Inter-American Institute for Global Change. A geographer, she studies the human dimensions of global change, with a focus on climate impacts, vulnerability and adaptation, and on the causes and consequences of climate change in Latin America. She is presently chair of the Science Advisory Committee for the international Global Environmental Change and Food Security programme.

Index